Principles of Materials Science and Engineering

Principles of Materials Science and Engineering

Edited by
Diane Jeffers

Willford Press
www.willfordpress.com

Published by Willford Press,
118-35 Queens Blvd., Suite 400,
Forest Hills, NY 11375, USA

ISBN: 978-1-68285-490-7

Cataloging-in-Publication Data

Principles of materials science and engineering / edited by Diane Jeffers.
p. cm.
Includes bibliographical references and index.
ISBN 978-1-68285-490-7
1. Materials science. 2. Materials. 3. Engineering. I. Jeffers, Diane.
TA403.6 .P75 2018
620.11--dc23

For information on all Willford Press publications
visit our website at www.willfordpress.com

Willford Press

Printed in China.

Contents

Preface

Materials science is an interdisciplinary subject, which is concerned with the design, discovery and study of new materials by using the elements of physics, chemistry, mineralogy, engineering, etc. The materials studied under this field are semiconductors, alloys, biomaterials, ceramic glasses, polymers, etc. The topics included in this book are of utmost significance and bound to provide incredible insights to readers. Those with an interest in the field of materials science and engineering would find this textbook helpful.

To facilitate a deeper understanding of the contents of this book a short introduction of every chapter is written below:

Chapter 1- Material science is an interdisciplinary field of engineering, physics and chemistry that studies the designing new materials. It is important to select the right material for a process. The chemical elements and its related processes determine the structure and function of a material. This chapter is an overview of the subject matter incorporating all the major aspects of material science.

Chapter 2- Materials can be normally classified into metals, ceramics and polymers, and each of them can be further subdivided into smaller sub-groups. Metals can be further simplified into ferrous and non-ferrous materials. Ceramics can be categories into refractory materials, glass, abrasives, and advanced ceramics. Materials are best understood in confluence with the major topics listed in the following section.

Chapter 3- Hardness is the most common mechanical property of a material used to classify and distinguish materials. It is the resistance of a material against any pressure that may lead to any deformation. This section elucidates the crucial theories and principles of the properties of materials and strengthening mechanisms.

Chapter 4- The state of matter when the physical properties of any material remain constant is called phase. The physical properties include density, magnetization, index of refraction, chemical composition etc. The phases of a material are studied using a phase diagram, which is a graph that charts the phases in relation to its pressure, temperature and volume. The major components of phase transformation in materials are discussed in this chapter.

Finally, I would like to thank the entire team involved in the inception of this book for their valuable time and contribution. This book would not have been possible without their efforts. I would also like to thank my friends and family for their constant support.

Editor

An Overview of Materials Science

Material science is an interdisciplinary field of engineering, physics and chemistry that studies the designing new materials. It is important to select the right material for a process. The chemical elements and its related processes determine the structure and function of a material. This chapter is an overview of the subject matter incorporating all the major aspects of material science.

Materials Science

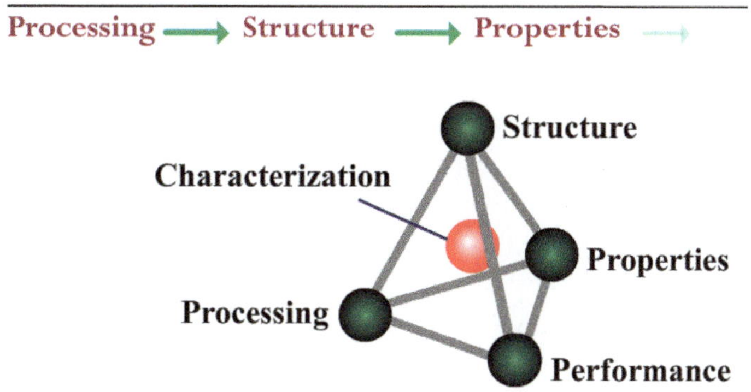

Material Science Tetrahedron

Material Science can be broadly defined as correlation between microstructure and properties.

The Materials Science Tetrahedron:-

Microstructure depends on the processing route while performance is dictated by properties.

Materials Science and Engineering (MSE)

In order to understand the structure of materials and its correlation to property, we have to start form the basic element of matter – The Atom

- An atom consists of a nucleus composed of protons and neutrons and electrons which encircle the nucleus.

- Protons and electrons have same and opposite charge of 1.6×10^{-19} C.
- Atomic number (Z) = Number protons = number of electrons.
- Atomic mass (A) = proton mass + neutron mass.
- Isotopes are the same element having different atomic masses. Number of protons in isotopes remains same while number of neutrons varies.

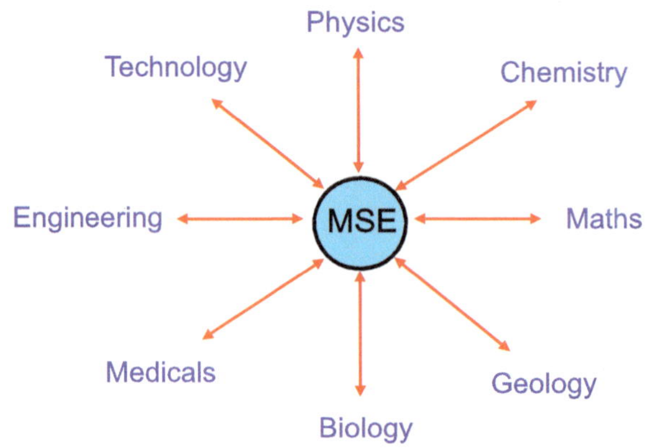

Materials Science is an interdisciplinary area where many science and engineering streams merge together

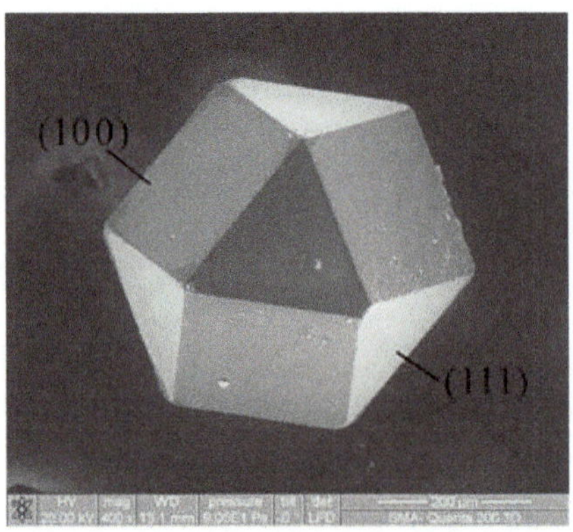

A diamond cuboctahedron showing seven crystallographic planes, imaged with scanning electron microscopy

The interdisciplinary field of materials science, also commonly termed materials science and engineering, involves the discovery and design of new materials, with an emphasis on solids. The intellectual origins of materials science stem from the Enlightenment, when researchers began to use analytical thinking from chemistry, physics, and engineering to understand ancient, phenomenological observations in metallurgy and mineralogy. Materials science still incorporates elements of physics, chemistry, and engineering. As such, the field was long considered by academic institutions as a subfield of these related fields. Beginning in the 1940s, materials science began to be more widely recognized as a specific and distinct field of science and engineering, and major technical universities around the world created dedicated schools of the study.

Materials science is a syncretic discipline hybridizing metallurgy, ceramics, solid-state physics, and chemistry. It is the first example of a new academic discipline emerging by fusion rather than fission.

Many of the most pressing scientific problems humans currently face are due to the limits of the materials that are available. Thus, breakthroughs in materials science are likely to affect the future of technology significantly.

Materials scientists emphasize understanding how the history of a material (its *processing*) influences its structure, and thus the material's properties and performance. The understanding of processing-structure-properties relationships is called the materials paradigm. This paradigm is used to advance understanding in a variety of research areas, including nanotechnology, biomaterials, and metallurgy. Materials science is also an important part of forensic engineering and failure analysis - investigating materials, products, structures or components which fail or which do not operate or function as intended, causing personal injury or damage to property. Such investigations are key to understanding, for example, the causes of various aviation accidents and incidents.

History

A late Bronze Age sword or dagger blade.

The material of choice of a given era is often a defining point. Phrases such as Stone Age, Bronze Age, Iron Age, and Steel Age are great examples. Originally deriving from the manufacture of ceramics and its putative derivative metallurgy, materials science is one of the oldest forms of engineering and applied science. Modern materials science evolved directly from metallurgy, which itself evolved from mining and (likely) ceramics and the use of fire. A major breakthrough in the understanding of materials occurred in the late 19th century, when the American scientist Josiah Willard Gibbs demonstrated that the thermodynamic properties related to atomic structure in various phases are related to the physical properties of a material. Important elements of modern materials science are a product of the space race: the understanding and engineering of the metallic alloys, and silica and carbon materials, used in building space vehicles enabling the exploration of space. Materials science has driven, and been driven by, the development of revolutionary technologies such as rubbers, plastics, semiconductors, and biomaterials.

Before the 1960s (and in some cases decades after), many *materials science* departments were named *metallurgy* departments, reflecting the 19th and early 20th century emphasis on metals. The growth of materials science in the United States was catalyzed in part by the Advanced Research Projects Agency, which funded a series of university-hosted laboratories in the early 1960s "to expand the national program of basic research and training in the materials sciences." The field has since broadened to include every class of materials, including ceramics, polymers, semiconductors, magnetic materials, medical implant materials, biological materials, and nanomaterials.

Fundamentals

A material is defined as a substance (most often a solid, but other condensed phases can be included) that is intended to be used for certain applications. There are a myriad of materials around us—they can be found in anything from buildings to spacecraft. Materials can generally be divided into two classes: crystalline and non-crystalline. The traditional examples of materials are metals, semiconductors, ceramics and polymers. New and advanced materials that are being developed include nanomaterials and biomaterials, etc.

The basis of materials science involves studying the structure of materials, and relating them to their properties. Once a materials scientist knows about this structure-property correlation, they can then go on to study the relative performance of a material in a given application. The major determinants of the structure of a material and thus of its properties are its constituent chemical elements and the way in which it has been processed into its final form. These characteristics, taken together and related through the laws of thermodynamics and kinetics, govern a material's microstructure, and thus its properties.

Structure

As mentioned above, structure is one of the most important components of the field of materials science. Materials science examines the structure of materials from the atomic scale, all the way up to the macro scale. Characterization is the way materials scientists examine the structure of a material. This involves methods such as diffraction with X-rays, electrons, or neutrons, and various forms of spectroscopy and chemical analysis such as Raman spectroscopy, energy-dispersive spectroscopy (EDS), chromatography, thermal analysis, electron microscope analysis, etc. Structure is studied at various levels, as detailed below.

Atomic Structure

This deals with the atoms of the materials, and how they are arranged to give molecules, crystals, etc. Much of the electrical, magnetic and chemical properties of materials arise from this level of structure. The length scales involved are in angstroms. The way in which the atoms and molecules are bonded and arranged is fundamental to studying the properties and behavior of any material.

Nanostructure

Buckminsterfullerene nanostructure.

Nanostructure deals with objects and structures that are in the 1—100 nm range. In many materials, atoms or molecules agglomerate together to form objects at the nanoscale. This causes many interesting electrical, magnetic, optical, and mechanical properties.

In describing nanostructures it is necessary to differentiate between the number of dimensions on the nanoscale. Nanotextured surfaces have *one dimension* on the nanoscale, i.e., only the thickness of the surface of an object is between 0.1 and 100 nm. Nanotubes have *two dimensions* on the nanoscale, i.e., the diameter of the tube is between 0.1 and 100 nm; its length could be much greater. Finally, spherical nanoparticles have *three dimensions* on the nanoscale, i.e., the particle is between 0.1 and 100 nm in each spatial dimension. The terms nanoparticles and ultrafine particles (UFP) often are used synonymously although UFP can reach into the micrometre range. The term 'nanostructure' is often used when referring to magnetic technology. Nanoscale structure in biology is often called ultrastructure.

Materials which atoms and molecules form constituents in the nanoscale (i.e., they form nanostructure) are called nanomaterials. Nanomaterials are subject of intense research in the materials science community due to the unique properties that they exhibit.

Microstructure

Microstructure is defined as the structure of a prepared surface or thin foil of material as revealed by a microscope above 25× magnification. It deals with objects from

100 nm to a few cm. The microstructure of a material (which can be broadly classified into metallic, polymeric, ceramic and composite) can strongly influence physical properties such as strength, toughness, ductility, hardness, corrosion resistance, high/low temperature behavior, wear resistance, and so on. Most of the traditional materials (such as metals and ceramics) are microstructured.

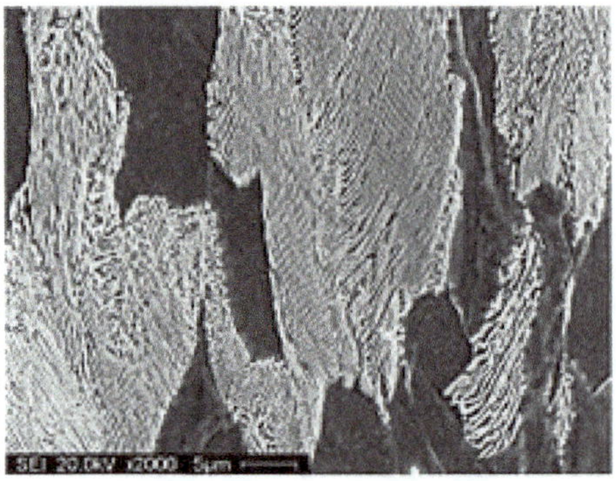

Microstructure of pearlite.

The manufacture of a perfect crystal of a material is physically impossible. For example, a crystalline material will contain defects such as precipitates, grain boundaries (Hall–Petch relationship), interstitial atoms, vacancies or substitutional atoms. The microstructure of materials reveals these defects, so that they can be studied.

Macro Structure

Macro structure is the appearance of a material in the scale millimeters to meters—it is the structure of the material as seen with the naked eye.

Crystallography

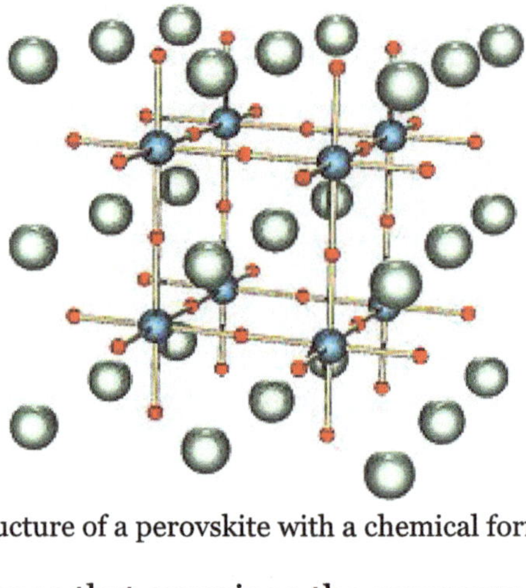

Crystal structure of a perovskite with a chemical formula ABX_3.

Crystallography is the science that examines the arrangement of atoms in crystalline solids. Crystallography is a useful tool for materials scientists. In single crystals, the effects of the crystalline arrangement of atoms is often easy to see macroscopically, because the natural shapes of crystals reflect the atomic structure. Further, physical properties are often controlled by crystalline defects. The understanding of crystal structures is an important prerequisite for understanding crystallographic defects. Mostly, materials do not occur as a single crystal, but in polycrystalline form, i.e., as an aggregate of small crystals with different orientations. Because of this, the powder diffraction method, which uses diffraction patterns of polycrystalline samples with a large number of crystals, plays an important role in structural determination. Most materials have a crystalline structure, but some important materials do not exhibit regular crystal structure. Polymers display varying degrees of crystallinity, and many are completely noncrystalline. Glass, some ceramics, and many natural materials are amorphous, not possessing any long-range order in their atomic arrangements. The study of polymers combines elements of chemical and statistical thermodynamics to give thermodynamic and mechanical, descriptions of physical properties.

Bonding

To obtain a full understanding of the material structure and how it relates to its properties, the materials scientist must study how the different atoms, ions and molecules are arranged and bonded to each other. This involves the study and use of quantum chemistry or quantum physics. Solid-state physics, solid-state chemistry and physical chemistry are also involved in the study of bonding and structure.

Properties

Materials exhibit myriad properties, including the following.

- Mechanical properties
- Chemical properties
- Electrical properties
- Thermal properties
- Optical properties
- Magnetic properties

The properties of a material determine its usability and hence its engineering application.

Synthesis and Processing

Synthesis and processing involves the creation of a material with the desired micro-nanostructure. From an engineering standpoint, a material cannot be used in industry if no economical production method for it has been developed. Thus, the processing of materials is vital to the field of materials science.

Different materials require different processing or synthesis methods. For example, the processing of metals has historically been very important and is studied under the branch of materials science named *physical metallurgy*. Also, chemical and physical methods are also used to synthesize other materials such as polymers, ceramics, thin films, etc. As of the early 21st century, new methods are being developed to synthesize nanomaterials such as graphene.

Thermodynamics

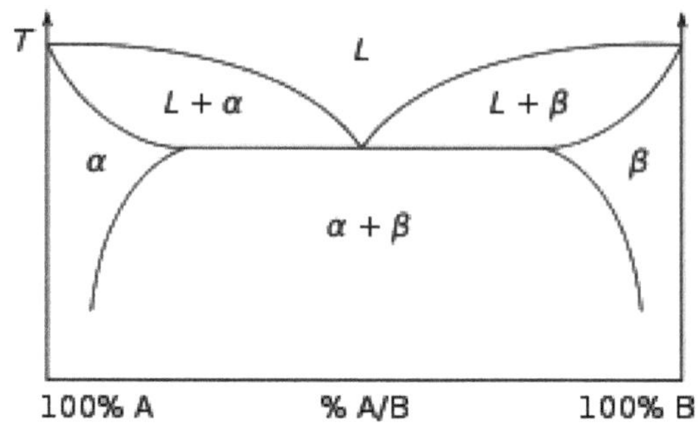

A phase diagram for a binary system displaying a eutectic point.

Thermodynamics is concerned with heat and temperature and their relation to energy and work. It defines macroscopic variables, such as internal energy, entropy, and pressure, that partly describe a body of matter or radiation. It states that the behavior of those variables is subject to general constraints, that are common to all materials, not the peculiar properties of particular materials. These general constraints are expressed in the four laws of thermodynamics. Thermodynamics describes the bulk behavior of the body, not the microscopic behaviors of the very large numbers of its microscopic constituents, such as molecules. The behavior of these microscopic particles is described by, and the laws of thermodynamics are derived from, statistical mechanics.

The study of thermodynamics is fundamental to materials science. It forms the foundation to treat general phenomena in materials science and engineering, including chemical reactions, magnetism, polarizability, and elasticity. It also helps in the understanding of phase diagrams and phase equilibrium.

Kinetics

Chemical kinetics is the study of the rates at which systems that are out of equilibrium change under the influence of various forces. When applied to materials science, it deals with how a material changes with time (moves from non-equilibrium to equilibrium state) due to application of a certain field. It details the rate of various processes evolving in materials including shape, size, composition and structure. Diffusion is important in the study of kinetics as this is the most common mechanism by which materials undergo change.

Kinetics is essential in processing of materials because, among other things, it details how the microstructure changes with application of heat.

In Research

Materials science has received much attention from researchers. In most universities, many departments ranging from physics to chemistry to chemical engineering, along with materials science departments, are involved in materials research. Research in materials science is vibrant and consists of many avenues. The following list is in no way exhaustive. It serves only to highlight certain important research areas.

Nanomaterials

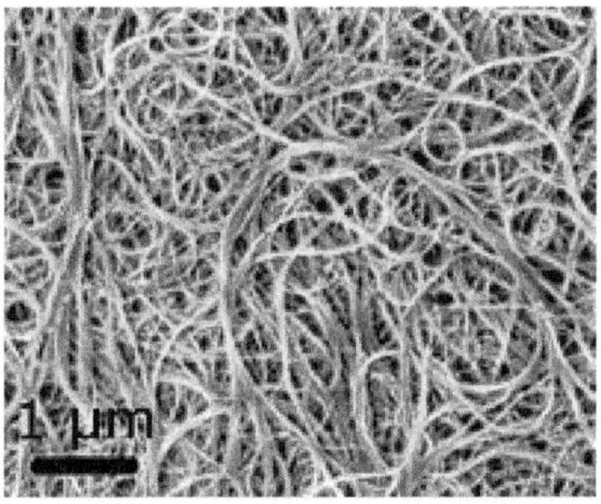

A scanning electron microscopy image of carbon nanotubes bundles

Nanomaterials describe, in principle, materials of which a single unit is sized (in at least one dimension) between 1 and 1000 nanometers (10^{-9} meter) but is usually 1—100 nm.

Nanomaterials research takes a materials science-based approach to nanotechnology, leveraging advances in materials metrology and synthesis which have been developed in support of microfabrication research. Materials with structure at the nanoscale often have unique optical, electronic, or mechanical properties.

The field of nanomaterials is loosely organized, like the traditional field of chemistry, into organic (carbon-based) nanomaterials such as fullerenes, and inorganic nanomaterials based on other elements, such as silicon. Examples of nanomaterials include fullerenes, carbon nanotubes, nanocrystals, etc.

Biomaterials

A biomaterial is any matter, surface, or construct that interacts with biological systems. As a science, *bio materials* is about fifty years old.The study of biomaterials is called *bio materials science*. It has experienced steady and strong growth over its history, with many companies investing large amounts of money into developing new products. Biomaterials science encompasses elements of medicine, biology, chemistry, tissue engineering, and materials science.

The iridescent nacre inside a nautilus shell.

Biomaterials can be derived either from nature or synthesized in a laboratory using a variety of chemical approaches using metallic components, polymers, bioceramics, or composite materials. They are often used and/or adapted for a medical application, and thus comprises whole or part of a living structure or biomedical device which performs, augments, or replaces a natural function. Such functions may be benign, like being used for a heart valve, or may be bioactive with a more interactive functionality such as hydroxylapatite coated hip implants. Biomaterials are also used every day in dental applications, surgery, and drug delivery. For example, a construct with impregnated pharmaceutical products can be placed into the body, which permits the prolonged release of a drug over an extended period of time. A biomaterial may also be an autograft, allograft or xenograft used as an organ transplant material.

Electronic, Optical, and Magnetic

Negative index metamaterial.

Semiconductors, metals, and ceramics are used today to form highly complex systems, such as integrated electronic circuits, optoelectronic devices, and magnetic and optical mass storage media. These materials form the basis of our modern computing world, and hence research into these materials is of vital importance.

Semiconductors are a traditional example of these types of materials. They are materials that have properties that are intermediate between conductors and insulators. Their electrical conductivities are very sensitive to impurity concentrations, and this allows for the use of doping to achieve desirable electronic properties. Hence, semiconductors form the basis of the traditional computer.

This field also includes new areas of research such as superconducting materials, spintronics, metamaterials, etc. The study of these materials involves knowledge of materials science and solid-state physics or condensed matter physics.

Computational Science and Theory

With the increase in computing power, simulating the behavior of materials has become possible. This enables materials scientists to discover properties of materials formerly unknown, as well as to design new materials. Up until now, new materials were found by time-consuming trial and error processes. But, now it is hoped that computational methods could drastically reduce that time, and allow tailoring materials properties. This involves simulating materials at all length scales, using methods such as density functional theory, molecular dynamics, etc.

In Industry

Radical materials advances can drive the creation of new products or even new industries, but stable industries also employ materials scientists to make incremental improvements and troubleshoot issues with currently used materials. Industrial applications of materials science include materials design, cost-benefit tradeoffs in industrial production of materials, processing methods (casting, rolling, welding, ion implantation, crystal growth, thin-film deposition, sintering, glassblowing, etc.), and analytic methods (characterization methods such as electron microscopy, X-ray diffraction, calorimetry, nuclear microscopy (HEFIB), Rutherford backscattering, neutron diffraction, small-angle X-ray scattering (SAXS), etc.).

Besides material characterization, the material scientist or engineer also deals with extracting materials and convering them into useful forms. Thus ingot casting, foundry methods, blast furnace extraction, and electrolytic extraction are all part of the required knowledge of a materials engineer. Often the presence, absence, or variation of minute quantities of secondary elements and compounds in a bulk material will greatly affect the final properties of the materials produced. For example, steels are classified based on 1/10 and 1/100 weight percentages of the carbon and other alloying elements they contain. Thus, the extracting and purifying methods used to extract iron in a blast furnace can affect the quality of steel that is produced.

Ceramics and Glasses

Another application of material science is the structures of ceramics and glass typically associated with the most brittle materials. Bonding in ceramics and glasses uses covalent and ionic-covalent types with SiO_2 (silica or sand) as a fundamental building block. Ceramics are as soft as clay or as hard as stone and concrete. Usually, they are crystalline in form. Most glasses contain a metal oxide fused with silica. At high temperatures used to prepare glass, the material is a viscous liquid. The structure of glass forms into an amorphous state upon cooling. Windowpanes and eyeglasses are important examples. Fibers of glass are also available. Scratch resistant Corning Gorilla Glass is a well-known example of the application of materials science to drastically improve the properties of common components. Diamond and carbon in its graphite form are considered to be ceramics.

Si_3N_4 ceramic bearing parts

Engineering ceramics are known for their stiffness and stability under high temperatures, compression and electrical stress. Alumina, silicon carbide, and tungsten carbide are made from a fine powder of their constituents in a process of sintering with a binder. Hot pressing provides higher density material. Chemical vapor deposition can place a film of a ceramic on another material. Cermets are ceramic particles containing some metals. The wear resistance of tools is derived from cemented carbides with the metal phase of cobalt and nickel typically added to modify properties.

Composites

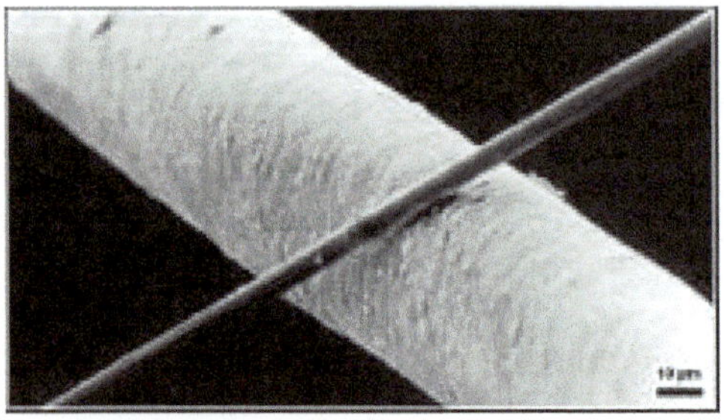

A 6 μm diameter carbon filament (running from bottom left to top right) siting atop the much larger human hair.

Filaments are commonly used for reinforcement in composite materials.

Another application of materials science in industry is making composite materials. These are structured materials composed of two or more macroscopic phases. Applications range from structural elements such as steel-reinforced concrete, to the thermal insulating tiles which play a key and integral role in NASA's Space Shuttle thermal protection system which is used to protect the surface of the shuttle from the heat of re-entry into the Earth's atmosphere. One example is reinforced Carbon-Carbon (RCC), the light gray material which withstands re-entry temperatures up to 1,510 °C (2,750 °F) and protects the Space Shuttle's wing leading edges and nose cap. RCC is a laminated composite material made from graphite rayon cloth and impregnated with a phenolic resin. After curing at high temperature in an autoclave, the laminate is pyrolized to convert the resin to carbon, impregnated with furfural alcohol in a vacuum chamber, and cured-pyrolized to convert the furfural alcohol to carbon. To provide oxidation resistance for reuse ability, the outer layers of the RCC are converted to silicon carbide.

Other examples can be seen in the "plastic" casings of television sets, cell-phones and so on. These plastic casings are usually a composite material made up of a thermoplastic matrix such as acrylonitrile butadiene styrene (ABS) in which calcium carbonate chalk, talc, glass fibers or carbon fibers have been added for added strength, bulk, or electrostatic dispersion. These additions may be termed reinforcing fibers, or dispersants, depending on their purpose.

Polymers

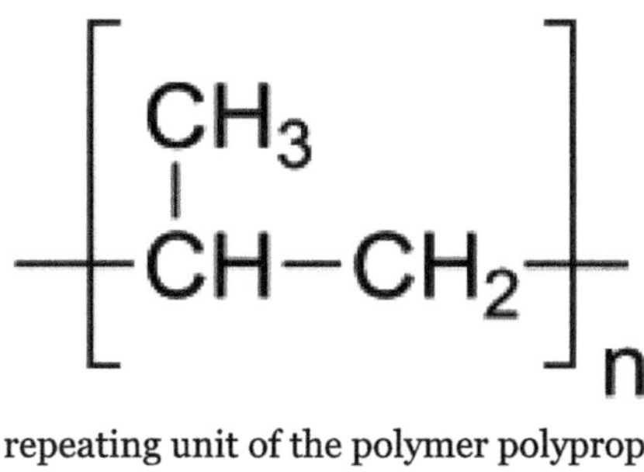

The repeating unit of the polymer polypropylene

Expanded polystyrene polymer packaging.

Polymers are chemical compounds made up of a large number of identical components linked together like chains. They are an important part of materials science. Polymers are the raw materials (the resins) used to make what are commonly called plastics and rubber. Plastics and rubber are really the final product, created after one or more polymers or additives have been added to a resin during processing, which is then shaped into a final form. Plastics which have been around, and which are in current widespread use, include polyethylene, polypropylene, polyvinyl chloride (PVC), polystyrene, nylons, polyesters, acrylics, polyurethanes, and polycarbonates and also rubbers which have been around are natural rubber, styrene butadiene rubber, chloroprene, and butadiene rubber. Plastics are generally classified as *commodity*, *specialty* and *engineering* plastics.

Polyvinyl chloride (PVC) is widely used, inexpensive, and annual production quantities are large. It lends itself to an vast array of applications, from artificial leather to electrical insulation and cabling, packaging, and containers. Its fabrication and processing are simple and well-established. The versatility of PVC is due to the wide range of plasticisers and other additives that it accepts. The term "additives" in polymer science refers to the chemicals and compounds added to the polymer base to modify its material properties.

Polycarbonate would be normally considered an engineering plastic (other examples include PEEK, ABS). Such plastics are valued for their superior strengths and other special material properties. They are usually not used for disposable applications, unlike commodity plastics.

Specialty plastics are materials with unique characteristics, such as ultra-high strength, electrical conductivity, electro-fluorescence, high thermal stability, etc.

The dividing lines between the various types of plastics is not based on material but rather on their properties and applications. For example, polyethylene (PE) is a cheap, low friction polymer commonly used to make disposable bags for shopping and trash, and is considered a commodity plastic, whereas medium-density polyethylene (MDPE) is used for underground gas and water pipes, and another variety called ultra-high-molecular-weight polyethylene (UHMWPE) is an engineering plastic which is used extensively as the glide rails for industrial equipment and the low-friction socket in implanted hip joints.

Metal Alloys

The study of metal alloys is a significant part of materials science. Of all the metallic alloys in use today, the alloys of iron (steel, stainless steel, cast iron, tool steel, alloy steels) make up the largest proportion both by quantity and commercial value. Iron alloyed with various proportions of carbon gives low, mid and high carbon steels. An iron-carbon alloy is only considered steel if the carbon level is between 0.01% and 2.00%. For the steels, the hardness and tensile strength of the steel is related to the amount of carbon present, with increasing carbon levels also leading to lower ductility and toughness. Heat treatment processes such as quenching and tempering can significantly change these properties, however. Cast Iron is defined as an iron–carbon alloy with more than 2.00% but less than 6.67% carbon. Stainless steel is defined as a regular steel alloy with greater than 10% by weight alloying content of Chromium. Nickel and Molybdenum are typically also found in stainless steels.

Wire rope made from steel alloy.

Other significant metallic alloys are those of aluminium, titanium, copper and magnesium. Copper alloys have been known for a long time (since the Bronze Age), while the alloys of the other three metals have been relatively recently developed. Due to the chemical reactivity of these metals, the electrolytic extraction processes required were only developed relatively recently. The alloys of aluminium, titanium and magnesium are also known and valued for their high strength-to-weight ratios and, in the case of magnesium, their ability to provide electromagnetic shielding. These materials are ideal for situations where high strength-to-weight ratios are more important than bulk cost, such as in the aerospace industry and certain automotive engineering applications.

Semiconductors

The study of semiconductors is a significant part of materials science. A semiconductor is a material that has a resistivity between a metal and insulator. Its electronic properties can be greatly altered through intentionally introducing impurities or doping. From these semiconductor materials, things such as diodes, transistors, light-emitting diodes (LEDs), and analog and digital electric circuits can be built, making them materials of interest in industry. Semiconductor devices have replaced thermionic devices (vacuum tubes) in most applications. Semiconductor devices are manufactured both as single discrete devices and as integrated circuits (ICs), which consist of a number—from a few to millions—of devices manufactured and interconnected on a single semiconductor substrate.

Of all the semiconductors in use today, silicon makes up the largest portion both by quantity and commercial value. Monocrystalline silicon is used to produce wafers used in the semiconductor and electronics industry. Second to silicon, gallium arsenide (GaAs) is the second most popular semiconductor used. Due to its higher electron mobility and saturation velocity compared to silicon, it is a material of choice for high-speed electronics applications. These superior properties are compelling reasons to use GaAs circuitry in mobile phones, satellite communications, microwave point-to-point links and higher frequency radar systems. Other semiconductor materials include germanium, silicon carbide, and gallium nitride and have various applications.

Relation to other Fields

Materials science evolved—starting from the 1960s—because it was recognized that to create, discover and design new materials, one had to approach it in a unified manner. Thus, materials science and engineering emerged at the intersection of various fields such as metallurgy, solid state physics, chemistry, chemical engineering, mechanical engineering and electrical engineering.

The field is inherently interdisciplinary, and the materials scientists/engineers must be aware and make use of the methods of the physicist, chemist and engineer. The field thus maintains close relationships with these fields. Also, many physicists, chemists and engineers also find themselves working in materials science.

The overlap between physics and materials science has led to the offshoot field of *materials physics*, which is concerned with the physical properties of materials. The approach is generally more macroscopic and applied than in condensed matter physics.

The field of materials science and engineering is important both from a scientific perspective, as well as from an engineering one. When discovering new materials, one encounters new phenomena that may not have been observed before. Hence, there is a lot of science to be discovered when working with materials. Materials science also provides a test for theories in condensed matter physics.

Materials are of the utmost importance for engineers, as the usage of the appropriate materials is crucial when designing systems. As a result, materials science is an increasingly important part of an engineer's education.

Atomic mass unit (amu) = 1/12 mass of Carbon 12 (^{12}C)

- 1 mol of substance contains 6.023×10^{23} (Avogadro's number) atoms or molecules.
- Atomic weight = 1 amu/atom (or molecule) = 1 g/mol = Wt. of 6.023×10^{23} atoms or molecules.
- For example, atomic weight of copper is 63.54 amu/atom or 63.54 g/mole

The Bohr Model

(a)

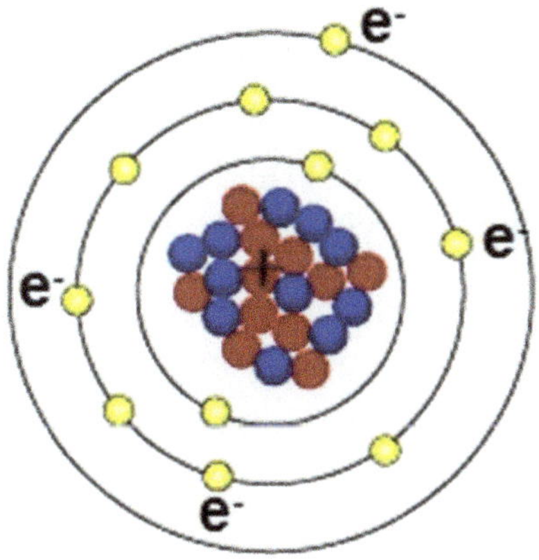

(b)

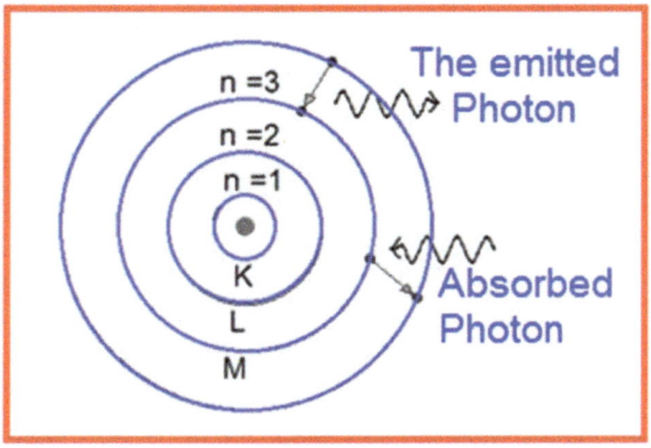

- Electrons revolve around a positively charged nucleus in discrete orbits (K, L, M or n=1, 2, 3 respectively) with specific levels of energy.
- Electrons positions are fixed as such, however, an electron can jump to higher or lower energy level by absorption or emission of energy respectively as shown in Figure (b)

Limitations of the Bohr Model

Although the Bohr's model was the first and best model available at the time of its discovery, it had certain limitations and could not explain many phenomena involving electrons.

Heisenberg's uncertainty principle:

The position and momentum of an electron can not be determined simultaneously.

This also disapproves the hypothesis in the Bohr model that electrons revolve around certain circular orbits.

Wave-Mechanical Model

- The wave-mechanical or wave-particle model was proposed to address the limitations in the Bohr model.
- The basic premise of this model is the wave-particle duality of electrons i.e. electrons are considered to have both wave-like and particle-like characteristics.
- The position of an electron is defined as the probability of finding it at different locations in an electron cloud around the nucleus i.e. position of an electron is described by a probability distribution instead of discrete orbits.

Wave-Mechanics Model

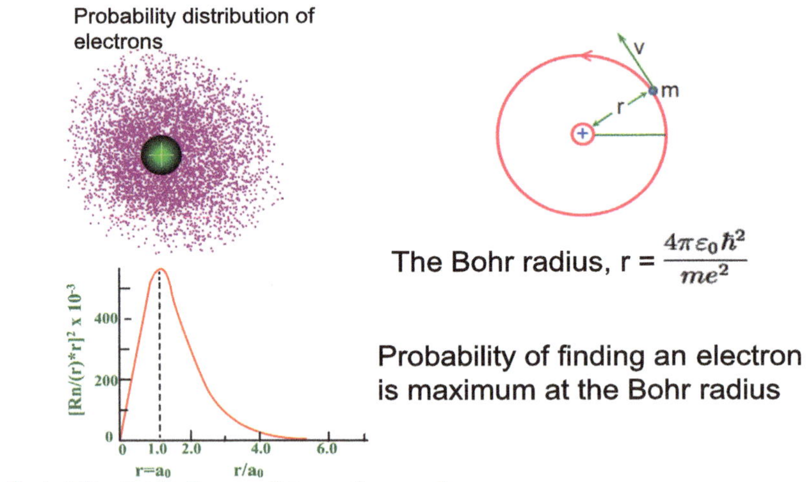

Probability distribution vs. distance from nucleus

Quantum Numbers

Four parameters or numbers called Quantum numbers are needed to describe the distribution and position of electrons in an atom.

The first three of them (n, ℓ, m_ℓ) describe the size, shape, and spatial orientation of the probability density distribution of electrons .

Principal Quantum Number, *n*

It describes electron shells as shown in the Bohr model. Values of n can be 1, 2, 3, 4 ... corresponding to electron shells K, L, M, N .....

The value of n also determines the size or distance of the shells from the nucleus.

Number of electrons in a shell = $2n^2$.e.g. number of electrons in K shell (n =1)= 2. 1^2 = 2, L shell (n =2) 8 (2. 2^2) and so on.

Atomic Bonding

(a)

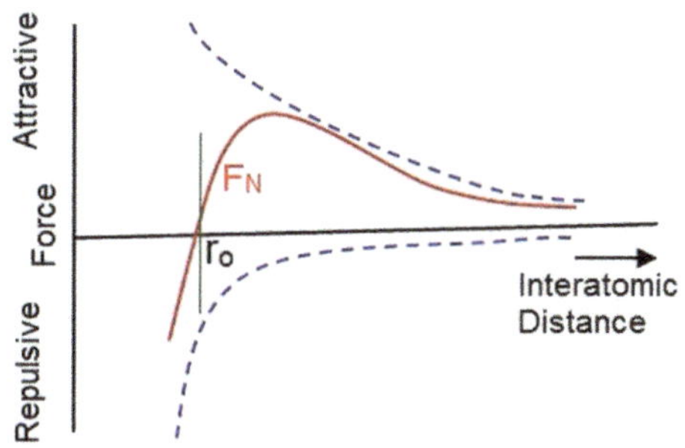

- When two neutral atoms are brought close to each other, they experience attractive and or repulsive force
- Attractive force is due to electrostatic attraction between electrons of one atom and the nucleus of the other.

Atomic Interaction

- Repulsive force arises due to repulsion between electrons and nuclei of the atoms.
- The net force, F_N (Fig. a), acting on the atoms is the summation of attractive and repulsive forces.
- The distance, at which the attraction and repulsion forces are equal and the net force is zero, is the equilibrium interatomic distance, r_o. The atoms have lowest energy at this position.
- Attraction is predominant above r_o and repulsion is dominant below r_o.

Lennard-Jones Potential

(b)

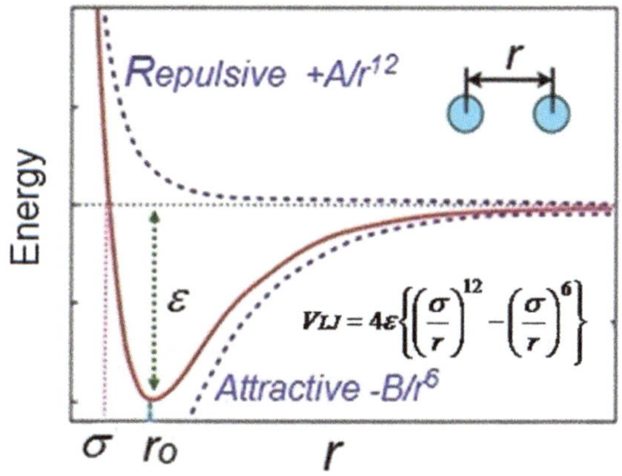

The interaction energy between the pair of atoms is given by the Lennard-Jones potential, V_r or $V_{LJ.}$

$$V_{LJ} = 4\varepsilon\left\{\left(\frac{\sigma}{r}\right)^{12} - \left(\frac{\sigma}{r}\right)^{6}\right\}$$

- σ is the distance at which the interaction energy is zero. ε is the depth of the potential well and is a measure of the bonding energy between two atoms.
- L-J potential can be also expressed in the simplified form as $V_{LJ} = A/r^{12} - B/r^6$ and hence, is also known as 6-12 potential.
- A/r^{12} is predominant at short distances and hence, represents the short-range repulsive potential due to overlap of electron orbitals and $-B/r^6$ is dominant at longer distance and hence, is the long range attractive potential.

Atomic Bonding

- The mechanisms of bonding between the atoms are based on the foregoing discussion on electrostatic inter- atomic interaction.
- The types of bond and bond strength are determined by the electronic structures of the atoms involved.
- The valence electrons take part in bonding. The atoms involved acquire, loose or share valence electrons to achieve the lowest energy or stable configuration of noble gases.
- Atomic bonding can be broadly classified as i) primary bonding ii) secondary bonding

Primary Bonds

Three types primary bonds are found in solids are:

- Ionic
- Covalent
- Mettalic

Majority of the engineering materials consist of one of these bonds. Many properties of the materials depend on the specific kind of bond and the bond energy.

Ionic Bond

Ionic bonds are generally found in compounds composed of metal and non-metal and arise out of electrostatic attraction between oppositely charged atoms (ions).

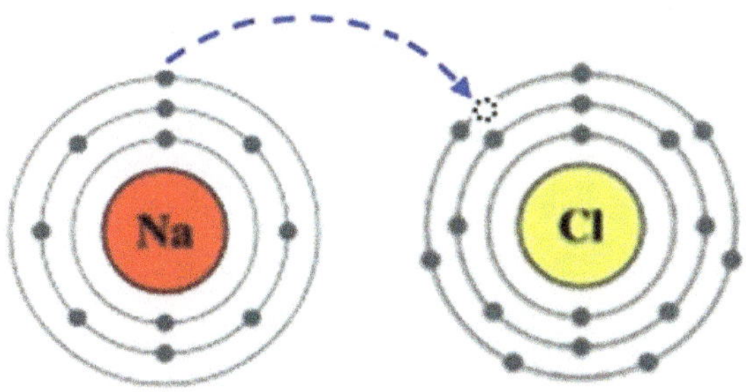

Number of electron in outer shell is 1 in Na and 7 in Cl . Therefore, Na will tend to reject one electron to get stable configuration of Ne and Cl will accept one electron to obtain Ar configuration. The columbic attraction between Na^+ and Cl^- ions thus formed will make an ionic bond to produce NaCl. Some other examples are CaF_2, CsCl , MgO, Al_2O_3.

CovalentBond

In this type of bonding, atoms share their valence electrons to get a stable configuration.

- Methane (CH_4): Four hydrogen atoms share their valence electrons with one carbon atom and the carbon atom in turn shares one valence electron with each of the four hydrogen atoms. In the process both H and C atoms get stable configuration and form a covalent bond.

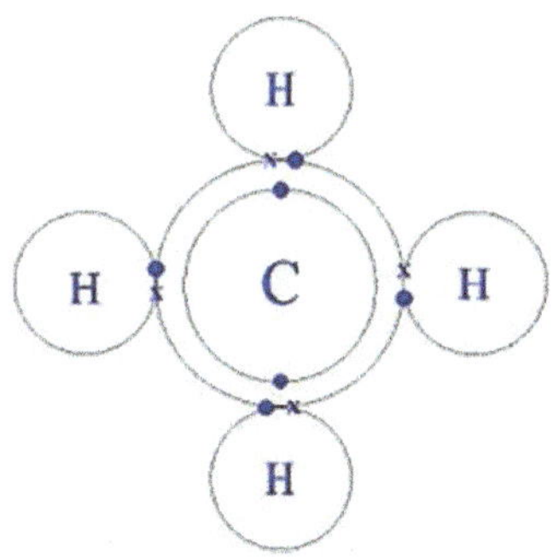

- Covalent bonds are formed between atoms of similar electronegativity.
- C atoms in diamond are covalently bonded to each other.
- Si also has valency of four and forms SiC through covalent bonding with C atoms.

Mettalic Bonds

In metals the valence electrons are not really bound to one particular atom, instead they form a sea or cloud of valence electrons which are shared by all the atoms. The remaining electrons and the nuclei form what is called the ion core which is positively charged. The metallic bond arises out of the columbic attraction between these two oppositely charged species – the electron cloud and the ion cores.

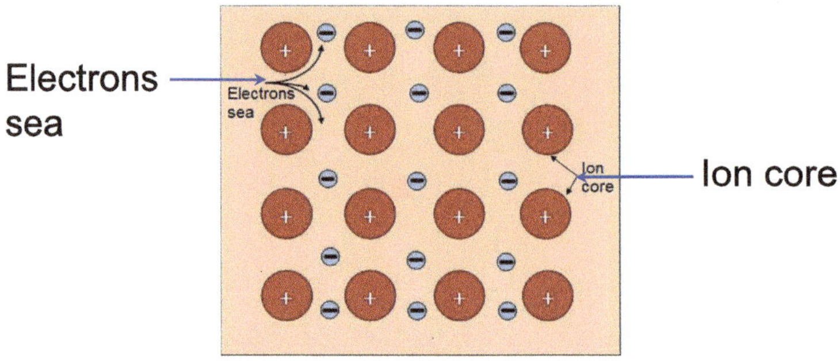

Characteristics of Primary Bonds

Structure-property Correlation

- Ionic and covalent bonds posses high bond energy – 450 – 1000 kJ/mole
- High bond strength in ionic and covalent solids results in high melting point, high strength and hardness. e.g. diamond
- As the electrons are tightly bound to the atoms they are generally poor conductors of heat and electricity
- Are brittle in nature
- Most of the ceramics consist of covalent (SiC) or ionic bonds (Al_2O_3) or a mix of both and hence, exhibit all the properties described above.
- Metallic bonds on the other hand provide good thermal and electrical conductivities as the valence electrons are free to move.
- The metallic bond energy is 68 kJ/mol (Hg) on the lower side and 850 kJ/mol (W, tungsten) on the higher side.
- Bond strength increases with atomic number as more electrons are available to form the bonds with the ion cores. As a result melting point, hardness and strength increases with atomic number.
- Metals are ductile as the free moving electrons provides agility to the bonds and allows plastic deformation.

Secondary Bonds

Van der Waals bonding

- Van der Waals bonding between molecules or atoms arise due to weak attraction forces between dipoles

- The natural oscillation of atoms leading to momentary break down of charge symmetry can generate temporary dipoles

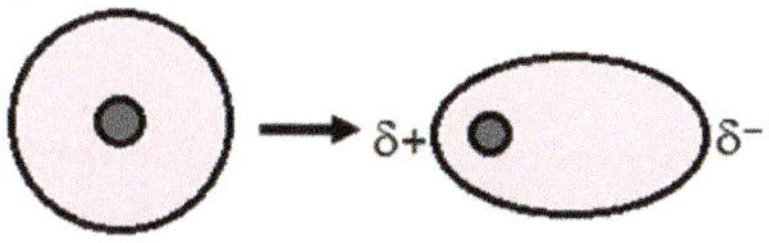

- Dipoles can induce dipoles and attraction between opposites ends of the dipoles leads to weak bonding

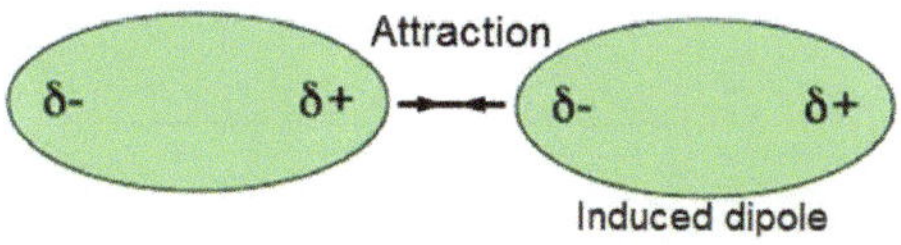

- An ion can also induce a dipole

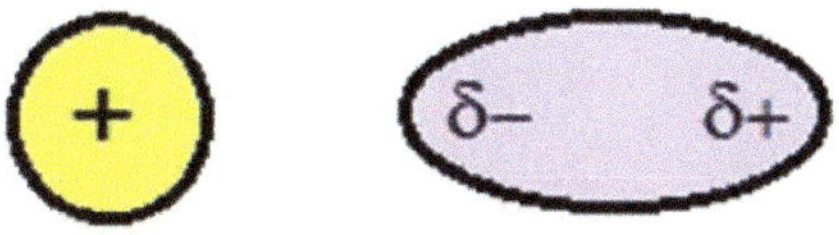

- Some molecules like HCl have permanent dipoles due to asymmetrical arrangement of +ve and –ve charges.

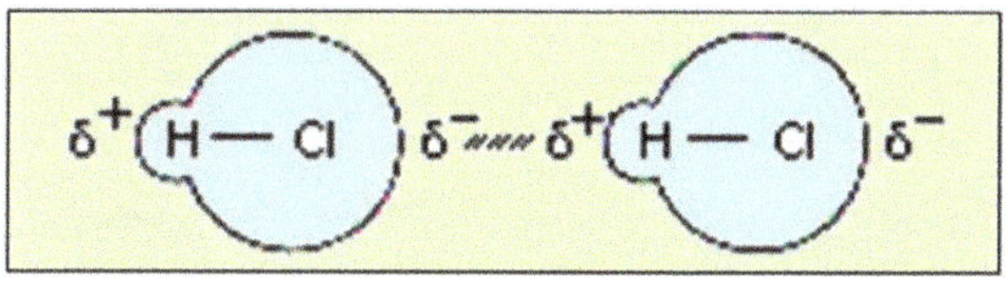

- Van der Waals bonding is much weaker compared to primary bonds. Bond energy lies in the range of 2 – 10 kJ/mol.
- Molecules in liquid and gas are held by weak Van der Waals forces
- The atomic layers in graphite are held together by weak van der Waals bonds. Therefore, the layers can move easily over each other and this imparts the lubricating property graphite is known for.

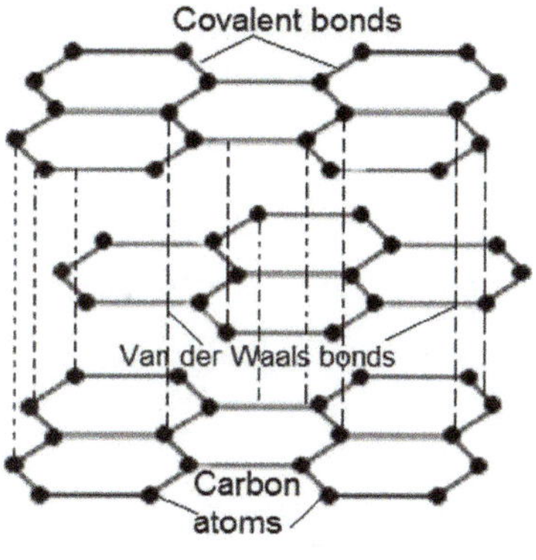

Hydrogen Bonding

- Hydrogen bond is a type of secondary bond found in molecules containing hydrogen as a constituent.
- The bond originates from electrostatic interaction between hydrogen and another atom of high electronegativity such as fluorine or oxygen.
- The strength of hydrogen bonds is in the range of 10 - 50 kJ/mol.
- Water molecules, for example, are connected by hydrogen bonds (dashed lines in the picture).

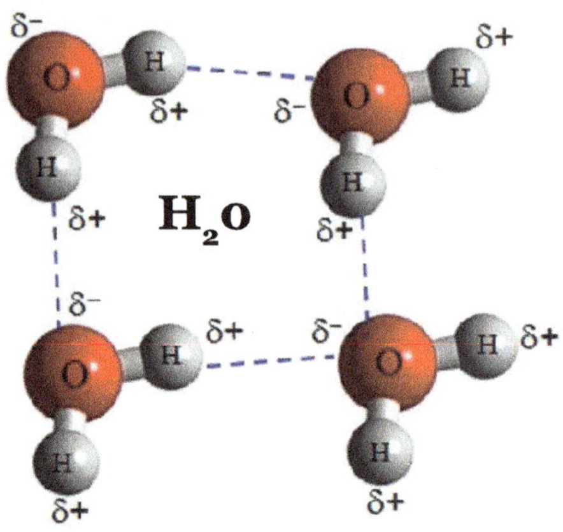

Crystal Structure

In crystallography, crystal structure is a description of the ordered arrangement of atoms, ions or molecules in a crystalline material. Ordered structures occur from the intrinsic nature of the constituent particles to form symmetric patterns that repeat along the principal directions of three-dimensional space in matter.

The smallest group of particles in the material that constitutes the repeating pattern is the unit cell of the structure. The unit cell completely defines the symmetry and structure of the entire crystal lattice, which is built up by repetitive translation of the unit cell along its principal axes. The repeating patterns are said to be located at the points of the Bravais lattice.

The lengths of the principal axes, or edges, of the unit cell and the angles between them are the lattice constants, also called *lattice parameters*. The symmetry properties of the crystal are described by the concept of space groups. All possible symmetric arrangements of particles in three-dimensional space may be described by the 230 space groups.

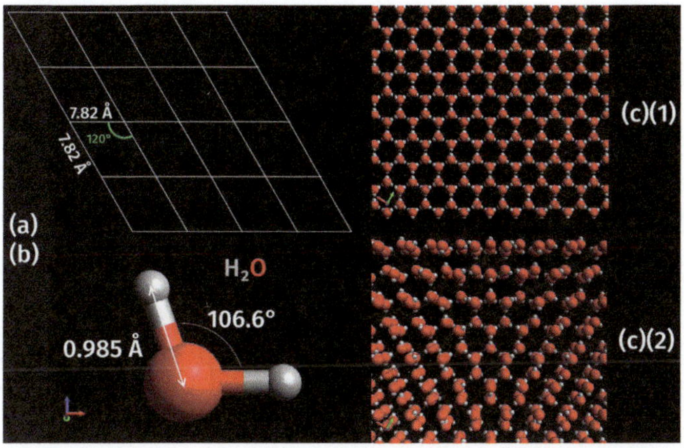

The (3-D) crystal structure of H_2O ice Ih (c) consists of bases of H_2O ice molecules (b) located on lattice points within the (2-D) hexagonal space lattice (a). The values for the H–O–H angle and O–H distance have come from *Physics of Ice* with uncertainties of ±1.5° and ±0.005 Å, respectively. The white box in (c) is the unit cell defined by Bernal and Fowler

The crystal structure and symmetry play a critical role in determining many physical properties, such as cleavage, electronic band structure, and optical transparency.

Unit Cell

The crystal structure of a material (the arrangement of atoms within a given type of crystal) can be described in terms of its unit cell. The unit cell is a box containing one or more atoms arranged in three dimensions. The unit cells stacked in three-dimensional space describe the bulk arrangement of atoms of the crystal. The unit cell is represented in terms of its lattice parameters, which are the lengths of the cell edges (a, b and c) and the angles between them (alpha, beta and gamma), while the positions of the atoms inside the unit cell are described by the set of atomic positions (x_i, y_i, z_i) measured from a reference lattice point. Commonly, atomic positions are represented in terms of fractional coordinates, relative to the unit cell lengths.

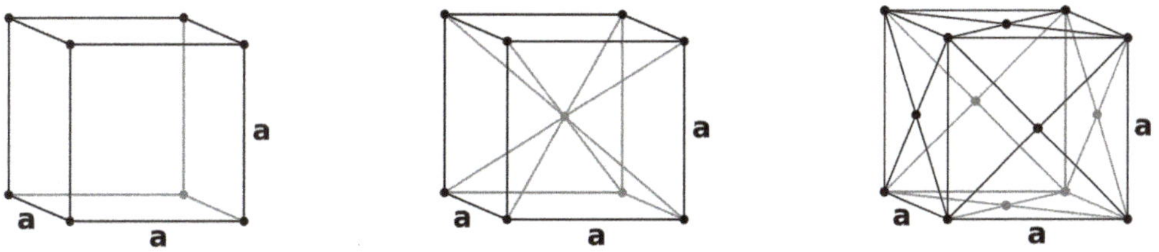

Simple cubic (P) Body-centered cubic (I) Face-centered cubic (F)

The atom positions within the unit cell can be calculated through application of symmetry operations to the asymmetric unit. The asymmetric unit refers to the smallest possible occupation of space within the unit cell. This does not, however imply that the entirety of the asymmetric unit must lie within the boundaries of the unit cell. Symmetric transformations of atom positions are calculated from the space group of the crystal structure, and this is usually a black box operation performed by computer programs. However, manual calculation of the atomic positions within the unit cell can be performed from the asymmetric unit, through the application of the symmetry operators described within the *International Tables for Crystallography: Volume A.*

Miller Indices

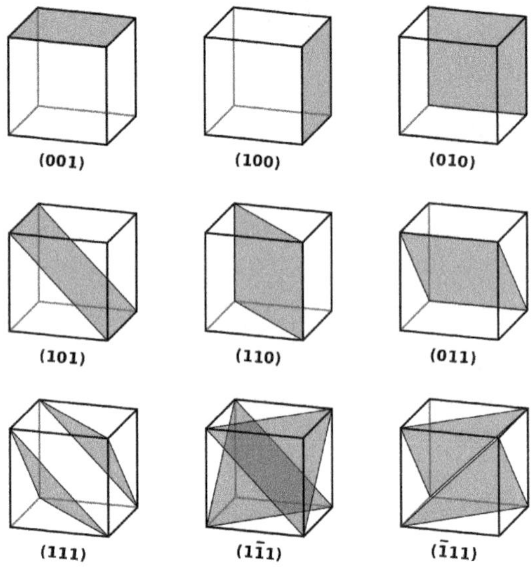

Planes with different Miller indices in cubic crystals

Vectors and planes in a crystal lattice are described by the three-value Miller index notation. It uses the indices ℓ, m, and n as directional parameters, which are separated by 90°, and are thus orthogonal.

By definition, the syntax (ℓmn) denotes a plane that intercepts the three points a_1/ℓ, a_2/m, and a_3/n, or some multiple thereof. That is, the Miller indices are proportional to the inverses of the intercepts of the plane with the unit cell (in the basis of the lattice vectors). If one or more of the indices is zero, it means that the planes do not intersect that axis (i.e., the intercept is "at infinity"). A plane containing a coordinate axis is translated so that it no longer contains that axis before its Miller indices are determined. The Miller indices for a plane are integers with no common factors. Negative indices are indicated with horizontal bars, as in (1$\bar{2}$3). In an orthogonal coordinate system for a cubic cell, the Miller indices of a plane are the Cartesian components of a vector normal to the plane.

Considering only (ℓmn) planes intersecting one or more lattice points (the *lattice planes*), the distance d between adjacent lattice planes is related to the (shortest) reciprocal lattice vector orthogonal to the planes by the formula.

$$d = \frac{2\pi}{|g_{\ell mn}|}$$

Planes and Directions

The crystallographic directions are geometric lines linking nodes (atoms, ions or molecules) of a crystal. Likewise, the crystallographic planes are geometric *planes* linking nodes. Some directions and planes have a higher density of nodes. These high density planes have an influence on the behavior of the crystal as follows:

- Optical properties: Refractive index is directly related to density (or periodic density fluctuations).
- Adsorption and reactivity: Physical adsorption and chemical reactions occur at or near surface atoms or molecules. These phenomena are thus sensitive to the density of nodes.
- Surface tension: The condensation of a material means that the atoms, ions or molecules are more stable if they are surrounded by other similar species. The surface tension of an interface thus varies according to the density on the surface.

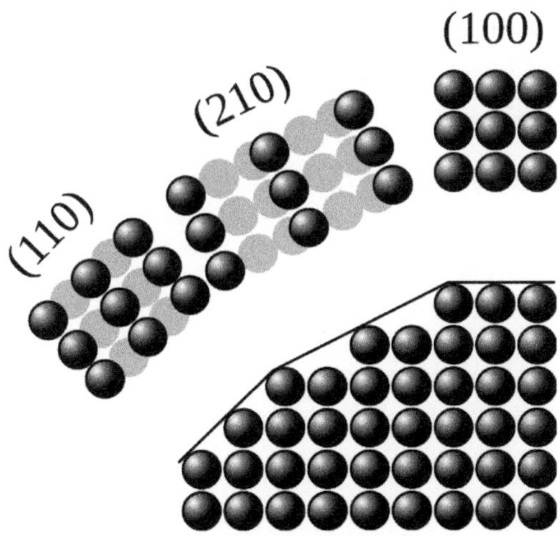

Dense crystallographic planes

- Microstructural defects: Pores and crystallites tend to have straight grain boundaries following higher density planes.
- Cleavage: This typically occurs preferentially parallel to higher density planes.
- Plastic deformation: Dislocation glide occurs preferentially parallel to higher density planes. The perturbation carried by the dislocation (Burgers vector) is along a dense direction. The shift of one node in a more dense direction requires a lesser distortion of the crystal lattice.

Some directions and planes are defined by symmetry of the crystal system. In monoclinic, rhombohedral, tetragonal, and trigonal/hexagonal systems there is one unique axis (sometimes called the principal axis) which has higher rotational symmetry than the other two axes. The basal plane is the plane perpendicular to the principal axis in these crystal systems. For triclinic, orthorhombic, and cubic crystal systems the axis designation is arbitrary and there is no principal axis.

Cubic Structures

For the special case of simple cubic crystals, the lattice vectors are orthogonal and of equal length (usually denoted *a*); similarly for the reciprocal lattice. So, in this common case, the Miller indices (*ℓmn*) and [*ℓmn*] both simply denote normals/directions in Cartesian coordinates. For cubic crystals with lattice constant *a*, the spacing *d* between adjacent (ℓmn) lattice planes is (from above):

$$d_{\ell mn} = \frac{a}{\sqrt{\ell^2 + m^2 + n^2}}$$

Because of the symmetry of cubic crystals, it is possible to change the place and sign of the integers and have equivalent directions and planes:

- Coordinates in *angle brackets* such as ⟨100⟩ denote a *family* of directions that are equivalent due to symmetry operations, such as [100], [010], [001] or the negative of any of those directions.
- Coordinates in *curly brackets* or *braces* such as {100} denote a family of plane normals that are equivalent due to symmetry operations, much the way angle brackets denote a family of directions.

For face-centered cubic (fcc) and body-centered cubic (bcc) lattices, the primitive lattice vectors are not orthogonal. However, in these cases the Miller indices are conventionally defined relative to the lattice vectors of the cubic supercell and hence are again simply the Cartesian directions.

Interplanar Spacing

The spacing *d* between adjacent (*hkl*) lattice planes is given by:

- Cubic:

$$\frac{1}{d^2} = \frac{h^2 + k^2 + l^2}{a^2}$$

- Tetragonal:

$$\frac{1}{d^2} = \frac{h^2 + k^2}{a^2} + \frac{l^2}{c^2}$$

- Hexagonal:

$$\frac{1}{d^2} = \frac{4}{3}\left(\frac{h^2 + hk + k^2}{a^2}\right) + \frac{l^2}{c^2}$$

- Rhombohedral:

$$\frac{1}{d^2} = \frac{(h^2 + k^2 + l^2)\sin^2\alpha + 2(hk + kl + hl)(\cos^2\alpha - \cos\alpha)}{a^2(1 - 3\cos^2\alpha + 2\cos^3\alpha)}$$

- Orthorhombic:

$$\frac{1}{d^2} = \frac{h^2}{a^2} + \frac{k^2}{b^2} + \frac{l^2}{c^2}$$

- Monoclinic:

$$\frac{1}{d^2} = \left(\frac{h^2}{a^2} + \frac{k^2\sin^2\beta}{b^2} + \frac{l^2}{c^2} - \frac{2hl\cos\beta}{ac}\right)\csc^2\beta$$

- Triclinic:

$$\frac{1}{d^2} = \frac{\frac{h^2}{a^2}\sin^2\alpha + \frac{k^2}{b^2}\sin^2\beta + \frac{l^2}{c^2}\sin^2\gamma + \frac{2kl}{bc}\cos\alpha + \frac{2hl}{ac}\cos\beta + \frac{2hk}{ab}\cos\gamma}{1 - \cos^2\alpha - \cos^2\beta - \cos^2\gamma + 2\cos\alpha\cos\beta\cos\gamma}$$

Classification by Symmetry

The defining property of a crystal is its inherent symmetry, by which we mean that under certain 'operations' the crystal remains unchanged. All crystals have translational symmetry in three directions, but some have other symmetry elements as well. For example, rotating the crystal 180° about a certain axis may result in an atomic configuration that is identical to the original configuration. The crystal is then said to have a twofold rotational symmetry about this axis. In addition to rotational symmetries like this, a crystal may have symmetries in the form of mirror planes and translational symmetries, and also the so-called "compound symmetries," which are a combination of translation and rotation/mirror symmetries. A full classification of a crystal is achieved when all of these inherent symmetries of the crystal are identified.

Lattice Systems

These lattice systems are a grouping of crystal structures according to the axial system used to describe their lattice. Each lattice system consists of a set of three axes in a particular geometric arrangement. There are seven lattice systems. They are similar to but not quite the same as the seven crystal systems.

Crystal family	Lattice system	Schönflies	14 Bravais Lattices			
			Primitive	Base-centered	Body-centered	Face-centered
triclinic		C_i	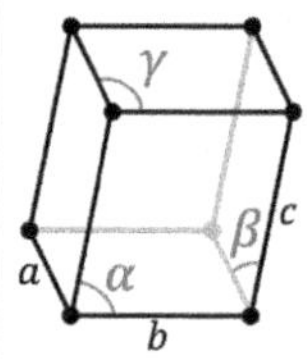			
monoclinic		C_{2h}	$\beta \neq 90°$ $a \neq c$ 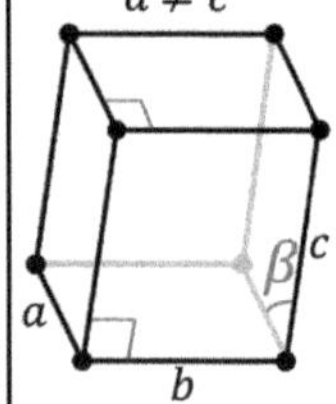	$\beta \neq 90°$ $a \neq c$ 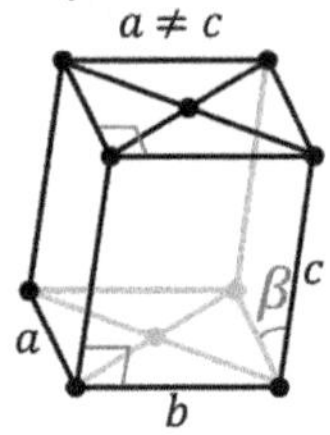		
orthorhombic		D_{2h}	$a \neq b \neq c$ 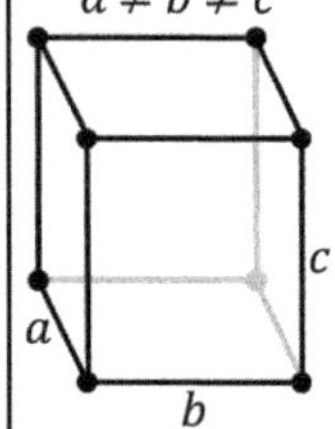	$a \neq b \neq c$ 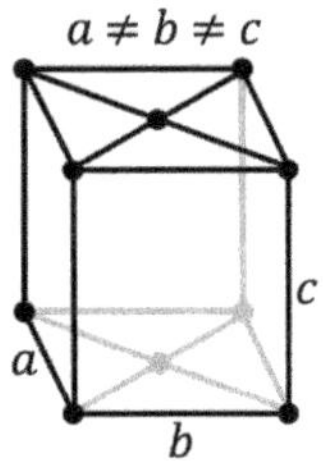	$a \neq b \neq c$ 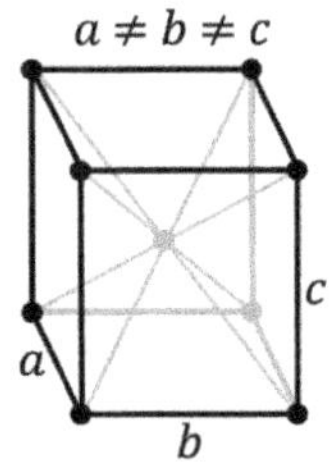	$a \neq b \neq c$ 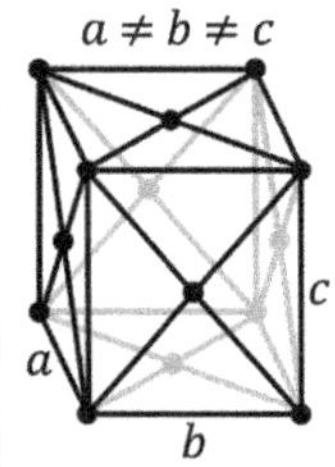
tetragonal		D_{4h}	$a \neq c$ 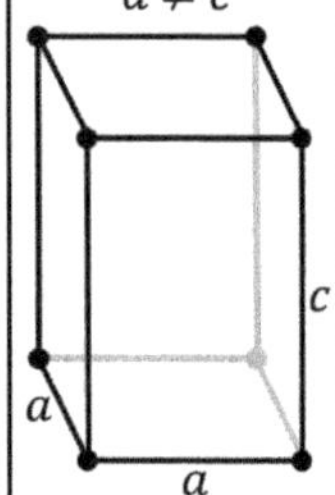		$a \neq c$ 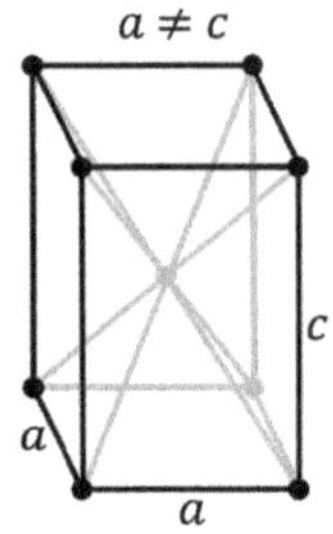	
hexagonal	rhombohedral	D_{3d}	$\alpha \neq 90°$ 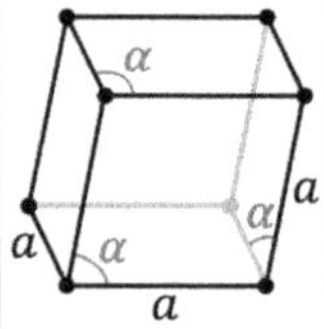			
	hexagonal	D_{6h}	$\gamma = 120°$ 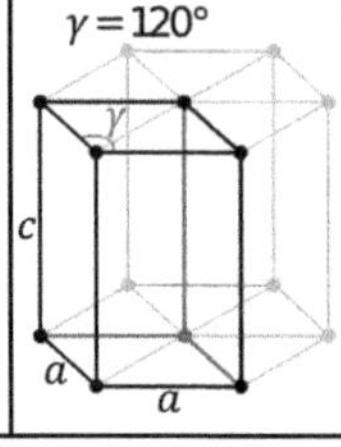			

cubic	O_h	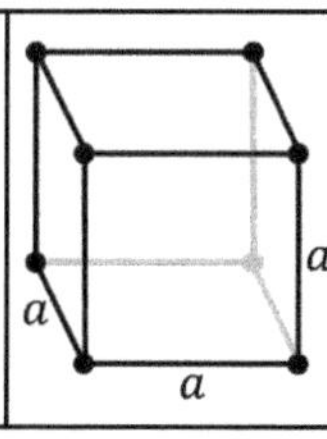		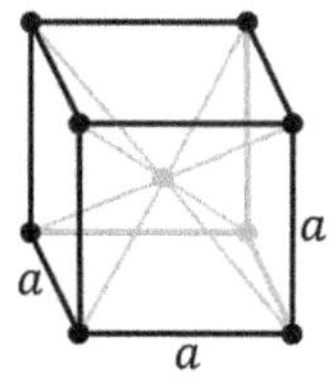	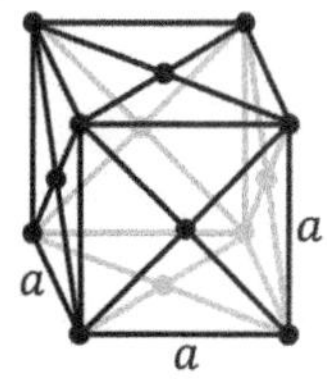

The simplest and most symmetric, the cubic (or isometric) system, has the symmetry of a cube, that is, it exhibits four threefold rotational axes oriented at 109.5° (the tetrahedral angle) with respect to each other. These threefold axes lie along the body diagonals of the cube. The other six lattice systems, are hexagonal, tetragonal, rhombohedral (often confused with the trigonal crystal system), orthorhombic, monoclinic and triclinic.

Bravais Lattices

Bravais lattices, also referred to as *space lattices*, describe the geometric arrangement of the lattice points, and therefore the translational symmetry of the crystal. The three dimensions of space afford 14 distinct Bravais lattices describing the translational symmetry. All crystalline materials recognized today, not including quasicrystals, fit in one of these arrangements. The fourteen three-dimensional lattices, classified by lattice system, are shown above.

The crystal structure consists of the same group of atoms, the *basis*, positioned around each and every lattice point. This group of atoms therefore repeats indefinitely in three dimensions according to the arrangement of one of the Bravais lattices. The characteristic rotation and mirror symmetries of the unit cell is described by its crystallographic point group.

Crystal Systems

A crystal system is a set of point groups in which the point groups themselves and their corresponding space groups are assigned to a lattice system. Of the 32 point groups that exist in three dimensions, most are assigned to only one lattice system, in which case the crystal system and lattice system both have the same name. However, five point groups are assigned to two lattice systems, rhombohedral and hexagonal, because both lattice systems exhibit threefold rotational symmetry. These point groups are assigned to the trigonal crystal system.

In total there are seven crystal systems: triclinic, monoclinic, orthorhombic, tetragonal, trigonal, hexagonal, and cubic.

Point Groups

The crystallographic point group or *crystal class* is the mathematical group comprising the symmetry operations that leave at least one point unmoved and that leave the appearance of the crystal structure unchanged. These symmetry operations include

- *Reflection*, which reflects the structure across a *reflection plane*
- *Rotation*, which rotates the structure a specified portion of a circle about a *rotation axis*
- *Inversion*, which changes the sign of the coordinate of each point with respect to a *center of symmetry* or *inversion point*
- *Improper rotation*, which consists of a rotation about an axis followed by an inversion.

Rotation axes (proper and improper), reflection planes, and centers of symmetry are collectively called *symmetry elements*. There are 32 possible crystal classes. Each one can be classified into one of the seven crystal systems.

Space Groups

In addition to the operations of the point group, the space group of the crystal structure contains translational symmetry operations. These include:

- Pure *translations*, which move a point along a vector
- *Screw axes*, which rotate a point around an axis while translating parallel to the axis.
- *Glide planes*, which reflect a point through a plane while translating it parallel to the plane.

There are 230 distinct space groups.

Atomic Coordination

By considering the arrangement of atoms relative to each other, their coordination numbers (or number of nearest neighbors), interatomic distances, types of bonding, etc., it is possible to form a general view of the structures and alternative ways of visualizing them.

Close Packing

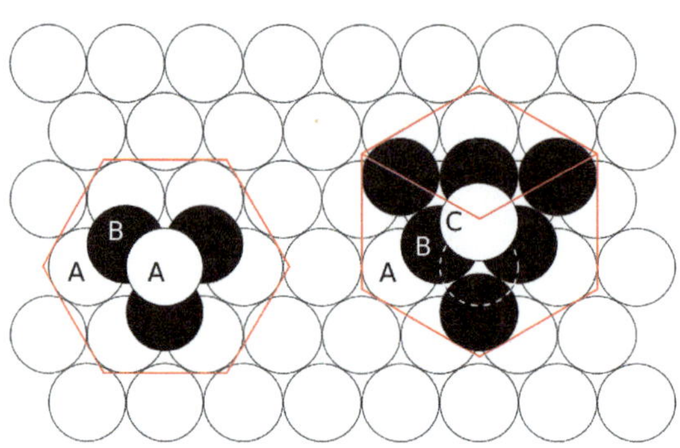

The hcp lattice (left) and the fcc lattice (right)

The principles involved can be understood by considering the most efficient way of packing together equal-sized spheres and stacking close-packed atomic planes in three dimensions. For example, if plane A lies beneath plane B, there are two possible ways of placing an additional atom on top of layer B. If an additional layer was placed directly over plane A, this would give rise to the following series:

...ABABABAB...

This arrangement of atoms in a crystal structure is known as hexagonal close packing (hcp).

If, however, all three planes are staggered relative to each other and it is not until the fourth layer is positioned directly over plane A that the sequence is repeated, then the following sequence arises:

...ABCABCABC...

This type of structural arrangement is known as cubic close packing (ccp).

The unit cell of a ccp arrangement of atoms is the face-centered cubic (fcc) unit cell. This is not immediately obvious as the closely packed layers are parallel to the {111} planes of the fcc unit cell. There are four different orientations of the close-packed layers.

The packing efficiency can be worked out by calculating the total volume of the spheres and dividing by the volume of the cell as follows:

$$\frac{4\times\frac{4}{3}\pi r^3}{16\sqrt{2}r^3}=\frac{\pi}{3\sqrt{2}}=0.7405...$$

The 74% packing efficiency is the maximum density possible in unit cells constructed of spheres of only one size. Most crystalline forms of metallic elements are hcp, fcc, or bcc (body-centered cubic). The coordination number of atoms in hcp and fcc structures is 12 and its atomic packing factor (APF) is the number mentioned above, 0.74. This can be compared to the APF of a bcc structure, which is 0.68.

Grain Boundaries

Grain boundaries are interfaces where crystals of different orientations meet. A grain boundary is a single-phase interface, with crystals on each side of the boundary being identical except in orientation. The term "crystallite boundary" is sometimes, though rarely, used. Grain boundary areas contain those atoms that have been perturbed from their original lattice sites, dislocations, and impurities that have migrated to the lower energy grain boundary.

Treating a grain boundary geometrically as an interface of a single crystal cut into two parts, one of which is rotated, we see that there are five variables required to define a grain boundary. The first two numbers come from the unit vector that specifies a rotation axis. The third number designates the angle of rotation of the grain. The final two numbers specify the plane of the grain boundary (or a unit vector that is normal to this plane).

Grain boundaries disrupt the motion of dislocations through a material, so reducing crystallite size is a common way to improve strength, as described by the Hall–Petch relationship. Since grain boundaries are defects in the crystal structure they tend to decrease the electrical and thermal conductivity of the material. The high interfacial energy and relatively weak bonding in most grain boundaries often makes them preferred sites for the onset of corrosion and for the precipitation of new phases from the solid. They are also important to many of the mechanisms of creep.

Grain boundaries are in general only a few nanometers wide. In common materials, crystallites are large enough that grain boundaries account for a small fraction of the material. However, very small grain sizes are achievable. In nanocrystalline solids, grain boundaries become a significant volume fraction of the material, with profound effects on such properties as diffusion and plasticity. In the limit of small crystallites, as the volume fraction of grain boundaries approaches 100%, the material ceases to have any crystalline character, and thus becomes an amorphous solid.

Defects and Impurities

Real crystals feature defects or irregularities in the ideal arrangements described above and it is these defects that critically determine many of the electrical and mechanical properties of real materials. When one atom substitutes for one of the principal atomic components within the crystal structure, alteration in the electrical and thermal properties of the material may ensue. Impurities may also manifest as spin impurities in certain materials. Research on magnetic impurities demonstrates that substantial alteration of certain properties such as specific heat may be affected by small concentrations of an impurity, as for example impurities in semiconducting ferromagnetic alloys may lead to different properties as first predicted in the late 1960s. Dislocations in the crystal lattice allow shear at lower stress than that needed for a perfect crystal structure.

Prediction of Structure

The difficulty of predicting stable crystal structures based on the knowledge of only the chemical composition has long been a stumbling block on the way to fully computational materials design. Now, with more powerful algorithms and high-performance computing, structures of medium complexity can be predicted using such approaches as evolutionary algorithms, random sampling, or metadynamics.

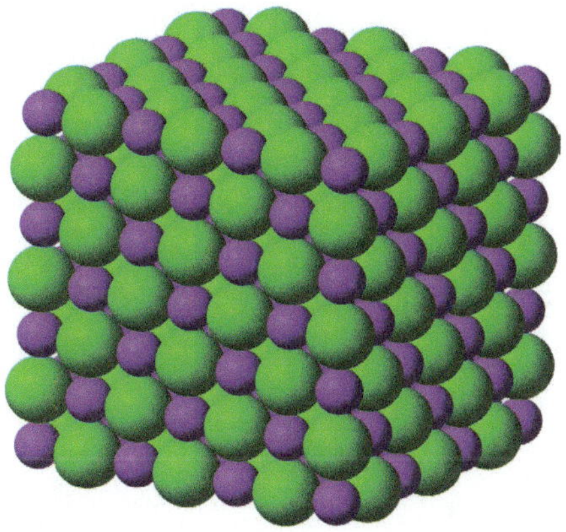

Crystal structure of sodium chloride (table salt)

The crystal structures of simple ionic solids (e.g., NaCl or table salt) have long been rationalized in terms of Pauling's rules, first set out in 1929 by Linus Pauling, referred to by many since as the "father of the chemical bond". Pauling also considered the nature of the interatomic forces in metals, and concluded that about half of the five d-orbitals in the transition metals are involved in bonding, with the remaining nonbonding d-orbitals being responsible for the magnetic properties. He, therefore, was able to correlate the number of d-orbitals in bond formation with the bond length as well as many of the physical properties of the substance. He subsequently introduced the metallic orbital, an extra orbital necessary to permit uninhibited resonance of valence bonds among various electronic structures.

In the resonating valence bond theory, the factors that determine the choice of one from among alternative crystal structures of a metal or intermetallic compound revolve around the energy of resonance of bonds among interatomic positions. It is clear that some modes of resonance would make larger contributions (be more mechanically stable than others), and that in particular a simple ratio of number of bonds to number of positions would be exceptional. The resulting principle is that a special stability is associated with the simplest ratios or "bond numbers": $1/2$, $1/3$, $2/3$, $1/4$, $3/4$, etc. The choice of structure and the value of the axial ratio (which determines the relative bond lengths) are thus a result of the effort of an atom to use its valency in the formation of stable bonds with simple fractional bond numbers.

After postulating a direct correlation between electron concentration and crystal structure in beta-phase alloys, Hume-Rothery analyzed the trends in melting points, compressibilities and bond lengths as a function of group number in the periodic table in order to establish a system of valencies of the transition elements in the metallic state. This treatment thus emphasized the increasing bond strength as a function of group number. The operation of directional forces were emphasized in one article on the relation between bond hybrids and the metallic structures. The resulting correlation between electronic and crystalline structures is summarized by a single parameter, the weight of the d-electrons per hybridized metallic orbital. The "d-weight" calculates out to 0.5, 0.7 and 0.9 for the fcc, hcp and bcc structures respectively. The relationship between d-electrons and crystal structure thus becomes apparent.

In crystal structure predictions/simulations, the periodicity is usually applied, since the system is imagined as unlimited big in all directions. Starting from a triclinic structure with no further symmetry property assumed, the system may be driven to show some additional symmetry properties by applying Newton's Second Law on particles in the unit cell and a recently developed dynamical equation for the system period vectors (lattice parameters including angles), even if the system is subject to external stress.

Polymorphism

Quartz is one of the several thermodynamically stable crystalline forms of silica, SiO_2. The most important forms of silica include: α-quartz, β-quartz, tridymite, cristobalite, coesite, and stishovite.

Polymorphism is the occurrence of multiple crystalline forms of a material. It is found in many crystalline materials including polymers, minerals, and metals. According to Gibbs' rules of phase equilibria, these unique crystalline phases are dependent on intensive variables such as pressure and temperature. Polymorphism is related to allotropy, which refers to elemental solids. The complete morphology of a material is described by polymorphism and other variables such as crystal habit, amorphous fraction or crystallographic defects. Polymorphs have different stabilities and may spontaneously convert from a metastable form (or thermodynamically unstable form) to the stable form at a particular temperature. They also exhibit different melting points, solubilities, and X-ray diffraction patterns.

One good example of this is the quartz form of silicon dioxide, or SiO_2. In the vast majority of silicates, the Si atom shows tetrahedral coordination by 4 oxygens. All but one of the crystalline forms involve tetrahedral $\{SiO_4\}$ units linked together by shared vertices in different arrangements. In different minerals the tetrahedra show different degrees of networking and polymerization. For example, they occur singly, joined together in pairs, in larger finite clusters including rings, in chains, double chains, sheets, and three-dimensional frameworks. The minerals are classified into groups based on these structures. In each of its 7 thermodynamically stable crystalline forms or polymorphs of crystalline quartz, only 2 out of 4 of each the edges of the $\{SiO_4\}$ tetrahedra are shared with others, yielding the net chemical formula for silica: SiO_2.

Another example is elemental tin (Sn), which is malleable near ambient temperatures but is brittle when cooled. This change in mechanical properties due to existence of its two major allotropes, α- and β-tin. The two allotropes that are encountered at normal pressure and temperature, α-tin and β-tin, are more commonly known as *gray tin* and *white tin* respectively. Two more allotropes, γ and σ, exist at temperatures above 161°C and pressures above several GPa. White tin is metallic, and is the stable crystalline form at or above room temperature. Below 13.2°C, tin exists in the gray form, which has a diamond cubic crystal structure, similar to diamond, silicon or germanium. Gray tin has no metallic properties at all, is a dull gray powdery material, and has few uses, other than a few specialized semiconductor applications. Although the α–β transformation temperature of tin is nominally 13.2°C, impurities (e.g. Al, Zn, etc.) lower the transition temperature well below 0°C, and upon addition of Sb or Bi the transformation may not occur at all.

Physical Properties

Twenty of the 32 crystal classes are piezoelectric, and crystals belonging to one of these classes (point groups) display piezoelectricity. All piezoelectric classes lack a center of symmetry. Any material develops a dielectric polarization when an electric field is applied, but a substance that has such a natural charge separation even in the absence of a field is called a polar material. Whether or not a material is polar is determined solely by its crystal structure. Only 10 of the 32 point groups are polar. All polar crystals are pyroelectric, so the 10 polar crystal classes are sometimes referred to as the pyroelectric classes.

There are a few crystal structures, notably the perovskite structure, which exhibit ferroelectric behavior. This is analogous to ferromagnetism, in that, in the absence of an electric field during production, the ferroelectric crystal does not exhibit a polarization. Upon the application of an electric field of sufficient magnitude, the crystal becomes permanently polarized. This polarization can be reversed by a sufficiently large counter-charge, in the same way that a ferromagnet can be reversed. However, although they are called ferroelectrics, the effect is due to the crystal structure (not the presence of a ferrous metal).

Crystal structure is one of the most important aspects of materials science and engineering as many properties of materials depend on their crystal structures. The basic principles of many materials characterization techniques such as X-ray diffraction (XRD), Transmission electron microscopy (TEM) are based on crystallography. Therefore, understanding the basics of crystal structures is of paramount importance.

Atomic Arrangement

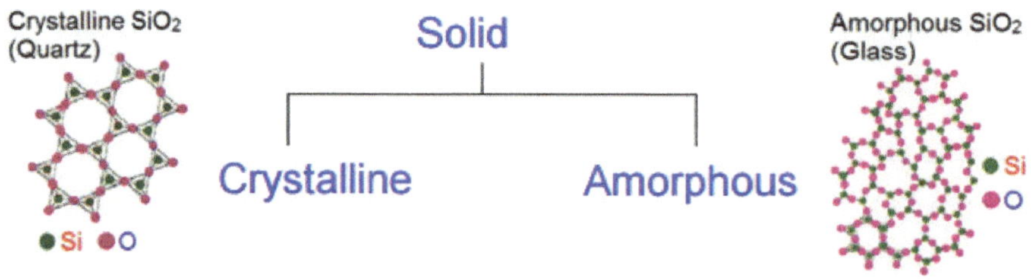

Crystalline – periodic arrangement of atoms: definite repetitive pattern.

Non-crystalline or Amorphous – random arrangement of atoms.

The periodicity of atoms in crystalline solids can be described by a network of points in space called lattice.

Space Lattice

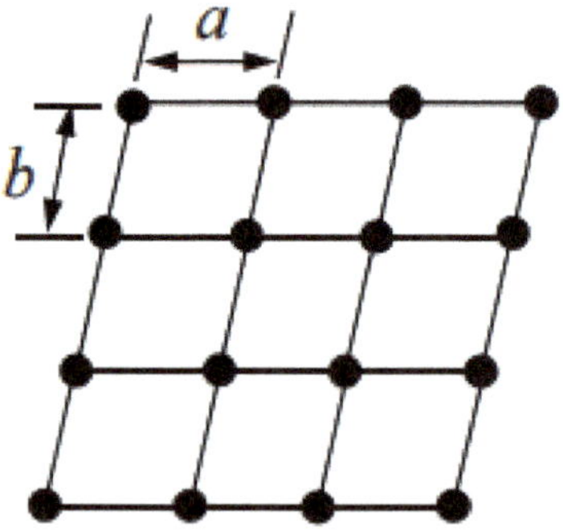

- A space lattice can be defined as a three dimensional array of points, each of which has identical surroundings.
- If the periodicity along a line is a, then position of any point along the line can be obtained by a simple translation, $r_u = ua$.
- Similarly $r_{uv} = ua + vb$ will repeat the point along a 2D plane, where u and v are integers.

Symmetry Operations

1. Translation
2. Rotation
3. Reflection
4. Inversion

Translation

The first point is repeated at equal distances along a line by a translation uT, where T is the translation vector and u is an integer.

Translation on a point with coordinates $xyz \rightarrow x+a \quad y+b \quad z+c$ where, a, b and c are the unit vectors in x, y and z directions respectively.

Rotation

A rotation can be applied on the translation vector T in all directions, clock or anti-clock wise, through equal angles α in the 2D space.

If two rotation operations, one each in clock and anti-clock direction, are applied on the translation vector T, it will create two more lattice points. Because of the regular pattern, the translation between these two points will be some multiple of T (pT).

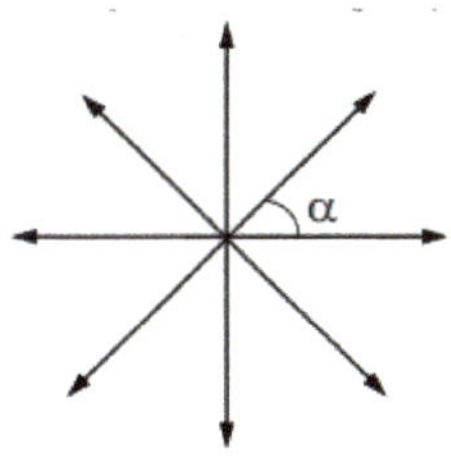

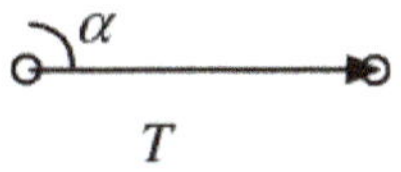

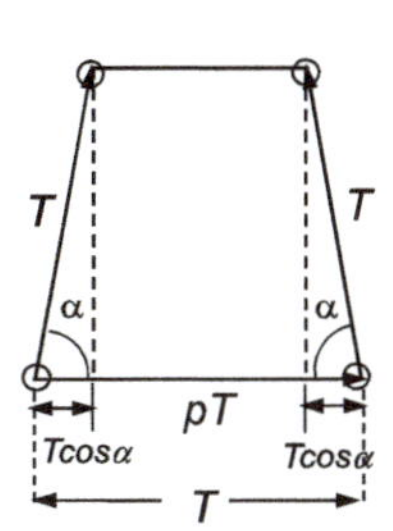

$$T = T\cos\alpha + pT + T\cos\alpha = pT + 2T\cos\alpha$$

$$\cos\alpha = (1 - p)/2$$

p	α°	n-fold	Symbol
0	60	6	⬢
1	90	4	■
2	120	3	▲
3	180	2	⬮
-1	0/360	1	

An n-fold rotation symmetry means rotation through an angle of $2\pi / n$ will repeat the object or motif n times in a full 360° rotation. n =1 means no symmetry.

1 Fold Rotation Axis - An object that requires rotation of a full 360° to repeat itself has no rotational symmetry.

2 Fold Rotation Axis - If an object appears identical after a rotation of 180°, that is twice in a 360° rotation, then it is said to have a 2-fold ($2\pi / 180$) rotation symmetry

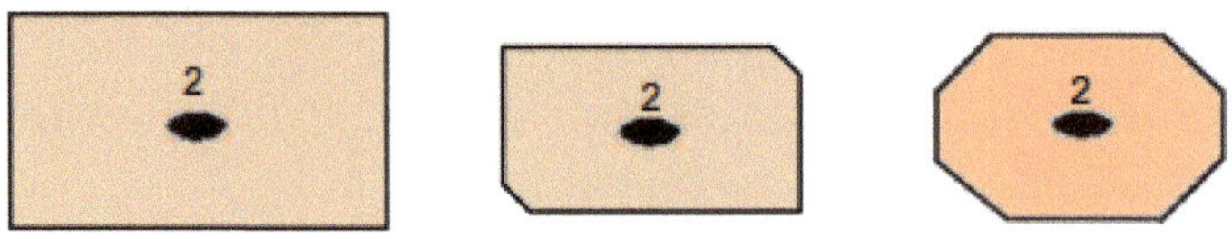

Similarly we have 3, 4 and 6-fold rotational symmetry

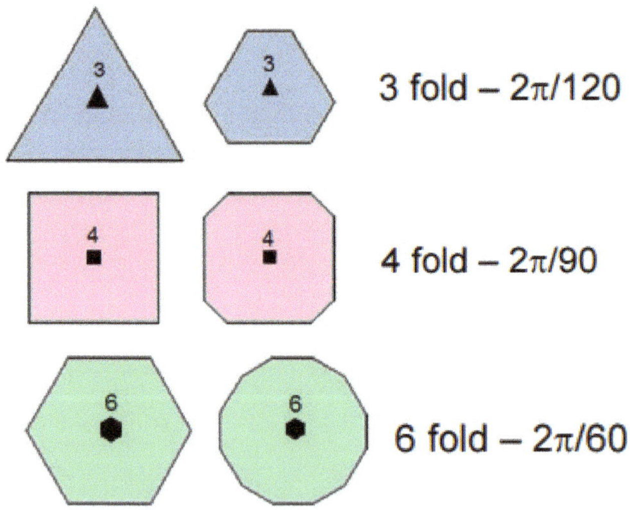

Is it possible to have 5, 7 or 8-fold rotation symmetry?

Objects with 5, 7 and 8 or higher order symmetry do exist in nature, e.g. star fish (5-fold), flowers with 5 or 8-fold symmetry.

However, these are not possible in crystallography as they cannot fill the space completely

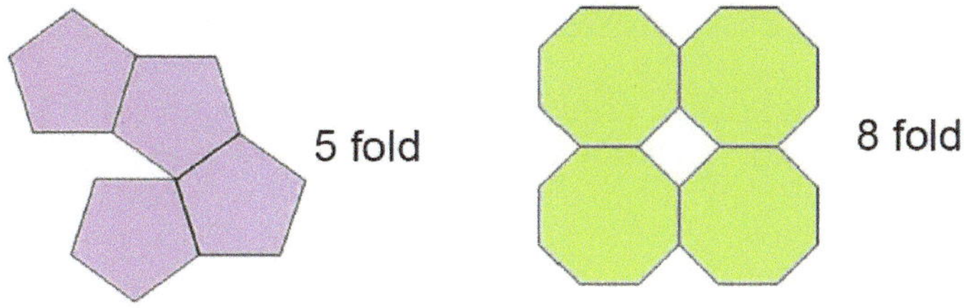

Reflection or Mirror symmetry

An object with a reflection symmetry will be a mirror image of itself across a plane called mirror plane (m).

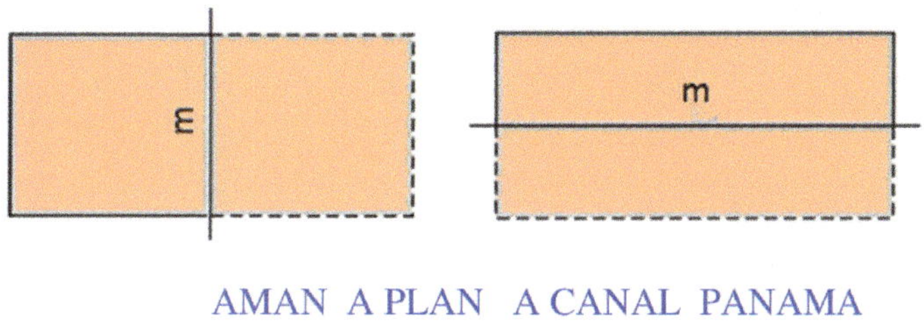

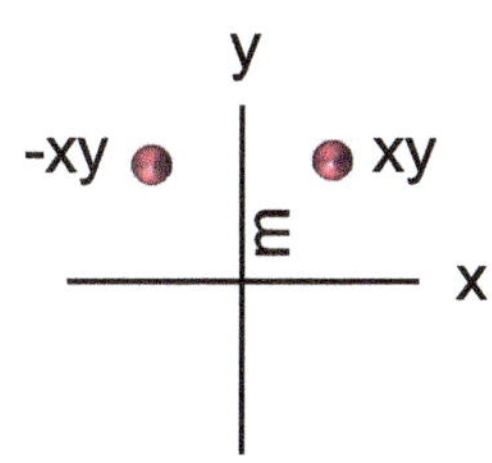

Reflection operation: $xyz \rightarrow -xyz$ $(\bar{x}\ y\ z)$ across a mirror plane perpendicular to x axis

Symmetry and Space Lattice

Symmetry elements discussed so far define five types of 2D space lattices. When a translation is applied to the third direction these lattices create a total of 7 crystal systems.

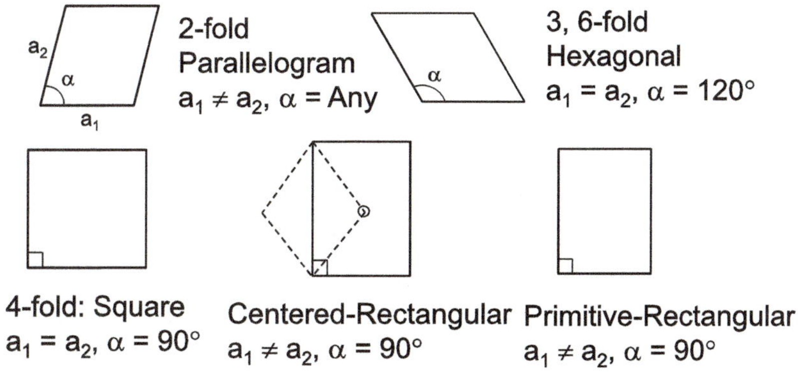

Inversion – Center of symmetry

In this operation, every part of the object is reflected through an inversion center called center of symmetry which is denoted as i. The object is reproduced inverted from its original position.

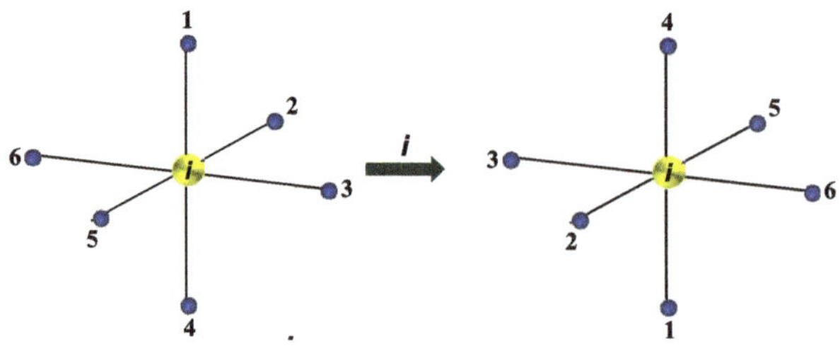

$$xyz \xrightarrow{i} -x\ -y\ -z\,(\bar{x}\,\bar{y}\,\bar{z})$$

Combined Operations

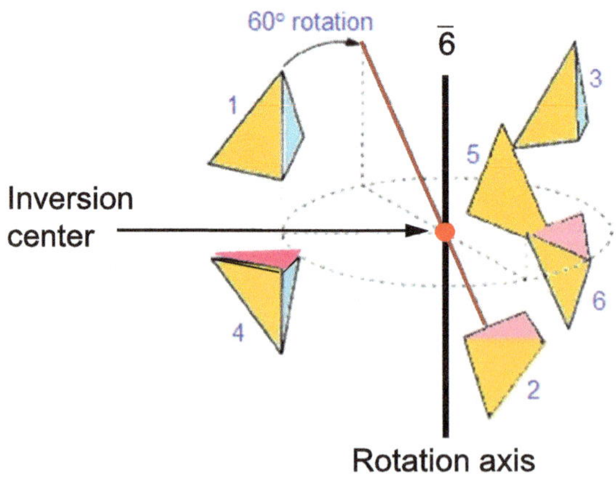

Combined symmetry operations also exist. For example, rotation can be combined with inversion which is called roto-inversion. The roto-inversion axis is denoted as $\bar{n}$. For example, a 6-fold roto-inversion ($\bar{6}$) involves rotating the object by 60° (360/6), and inverting through a symmetry center.

Point and Space groups

Symmetry operations generate a variety of arrangements of lattice points in three dimensions. There are 32 unique ways in which lattice points can be arranged in space. These non-translation elements are called point-groups.

A large number of 3D structures are generated when translations [linear translation, translation + reflection (glide plane) and translation + rotation (screw axis)] are applied to the point groups. There are 230 unique shapes which can be generated this way. These are called space groups.

Hermann-Mauguin Symbols

- The 32 point groups are denoted by notations called Hermann-Mauguin symbols. These symbols basically describe the unique symmetry elements present in a body.
- The shape in Fig.(a) contains 1 4-fold axis, 4 2-fold axes, 5 mirror planes. 3 mirror planes and 2 2-fold axes are unique as others can be produced by a symmetry operation. Therefore, point group symbol for this shape is 4/m2/m2/m. The "/" between 4 or 2 and m indicates that they are perpendicular to each other.

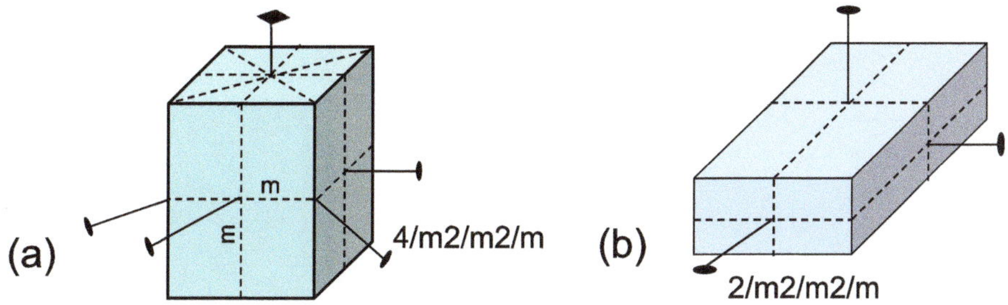

- Space lattice is arrangement of points with each point having exactly same surroundings.
- Symmetry operations restore a body to its original position.
- There are four symmetry operations – Translation, reflection, rotation and inversion.
- There are 32 point groups and 230 space groups.
- Hermann-Mauguin symbols are used to denote point groups.

Crystal Systems

- The space lattice points in a crystal are occupied by atoms.
- The position of any atom in the 3D lattice can be described by a vector $r_{uvw} = ua + vb + wc$, where u, v and w are integers.

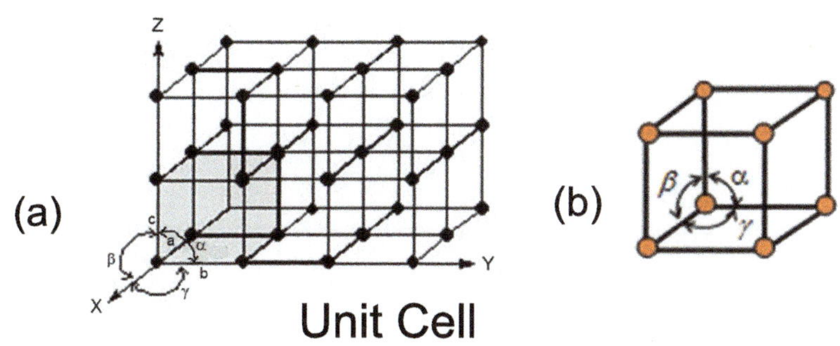

The three unit vectors, a, b, c can define a cell as shown by the shaded region in Fig. (a) This cell is known as unit cell (Fig. b) which when repeated in the three dimensions generates the crystal structure.

Bravais Lattice

The unit vectors *a*, *b* and *c* are called lattice parameters. Based on their length equality or inequality and their orientation (the angles between them, α, β and γ) a total of 7 crystal systems can be defined. With the centering (face, base and body centering) added to these, 14 kinds of 3D lattices, known as *Bravais lattices*, can be generated.

Cubic: $a = b = c$, $\alpha = \beta = \gamma = 90^\circ$

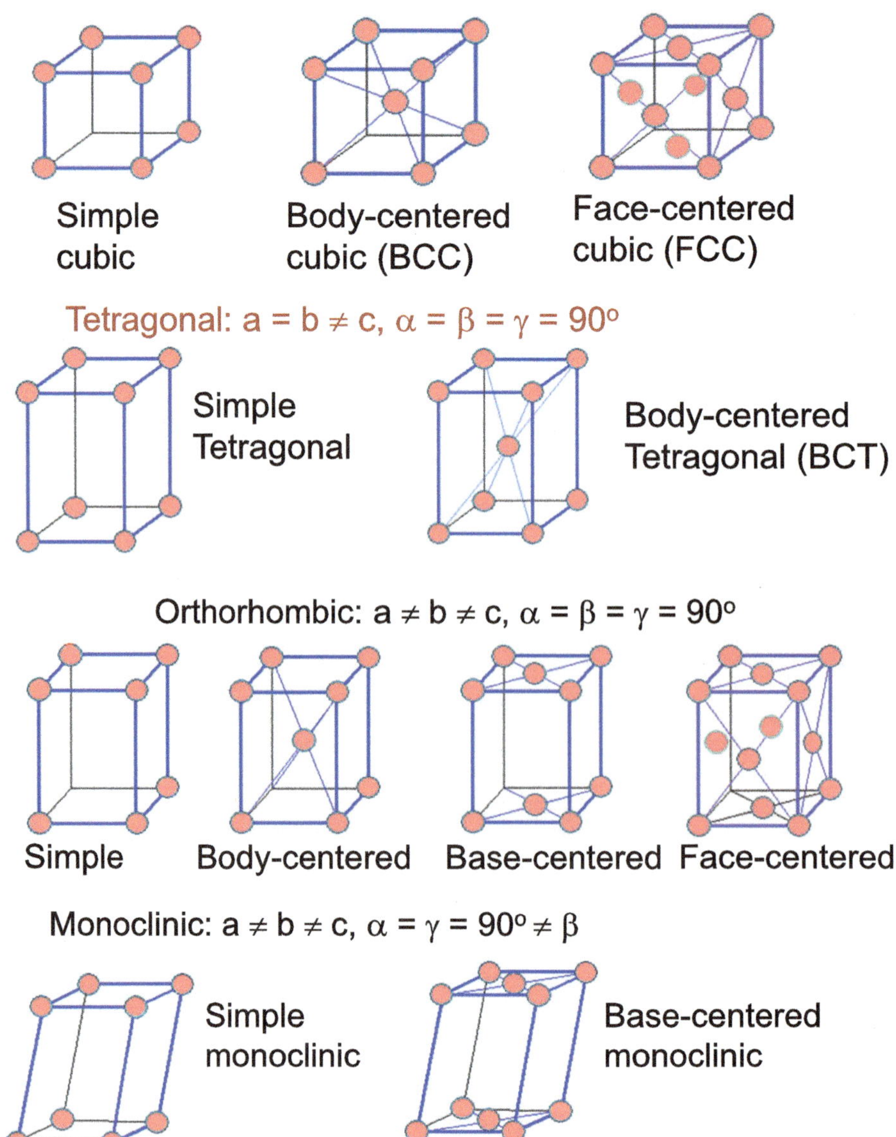

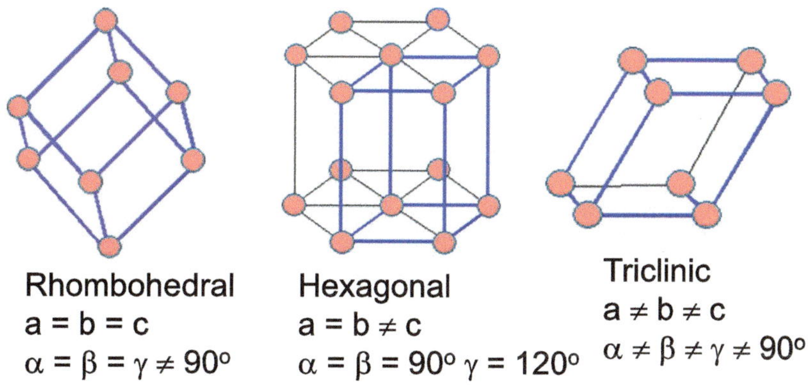

Crystal system	**Example**
Triclinic	$K_2S_2O_8$,$K_2Cr_2O_7$
Monoclinic	As_4S_4, KNO_2,$CaSO_4.2H_2O$, *β*-S
Rhombohedral	Hg, Sb, As, Bi, $CaCO_3$
Hexagonal	Zn, Co, Cd, Mg, Zr, NiAs
Orthorhombic	Ga, Fe_3C, *α*-S
Tetragonal	In, TiO_2, *β*-Sn
Cubic	Au, Si, Al, Cu, Ag, Fe, NaCl

Point Coordinates

Position of any point in a unit cell is given by its coordinates or distances from the x, y and z axes in terms of the lattice vectors a, b and c.

Thus the point located at a/2 along x axis, b/3 along y axis and c/2 along z axis, as shown in the figure below, has the coordinates $\frac{1}{2},\frac{1}{3},\frac{1}{2}$.

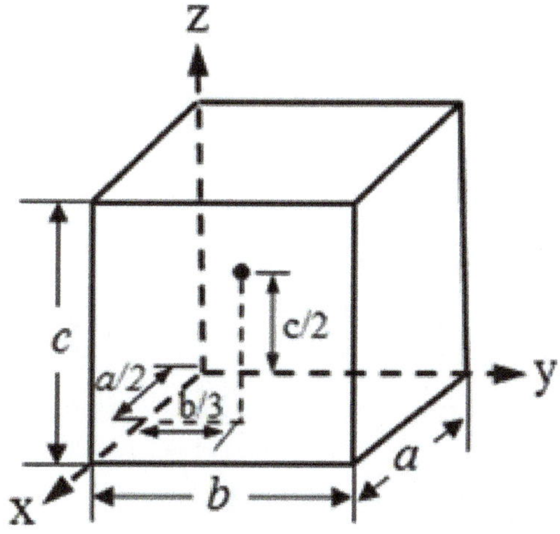

Crystal Planes

Miller Indices

Planes in a crystal are described by notations called Miller indices

- Miller indices of a plane, indicated by h k l, are given by the reciprocal of the intercepts of the plane on the three axes.
- The plane, which intersects X axis at 1 (one lattice parameter) and is parallel to Y and Z axes, has Miller indices $h = 1/1 = 1$, $k = 1/\infty = 0$, $l = 1/\infty = 0$. It is written as (hkl) = (100).
- Miller indices of some other planes in the cubic system are shown in the figures.

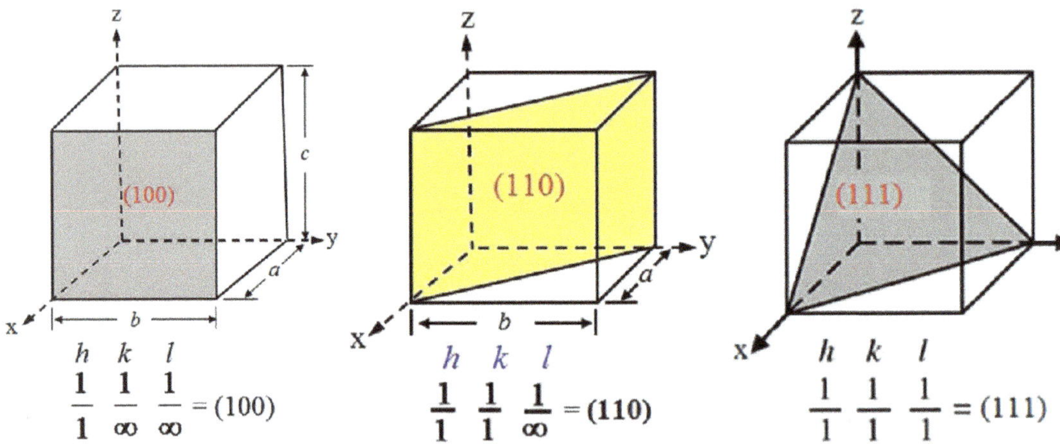

To find the Miller Indices of a plane, follow these steps:

- Determine the intercepts of the plane along the crystal axes
- Take the reciprocals
- Clear fractions
- Reduce to lowest terms and enclose in brackets ()

Ex: Intercepts on a, b, c : ¾, ½, ¼ (h k l) = (4/3, 2, 4) = (2 3 6)

Planes can also have negative intercept e.g. 1, -1/2, 1 h k l = 1 -2 1. This is denoted as $(1\bar{2}1)$

Family of Planes {hkl}

Planes having similar indices are equivalent, e.g. faces of the cube (100), (010) and (001). This is termed as a family of planes and denoted as {100} which includes all the (100) combinations including negative indices.

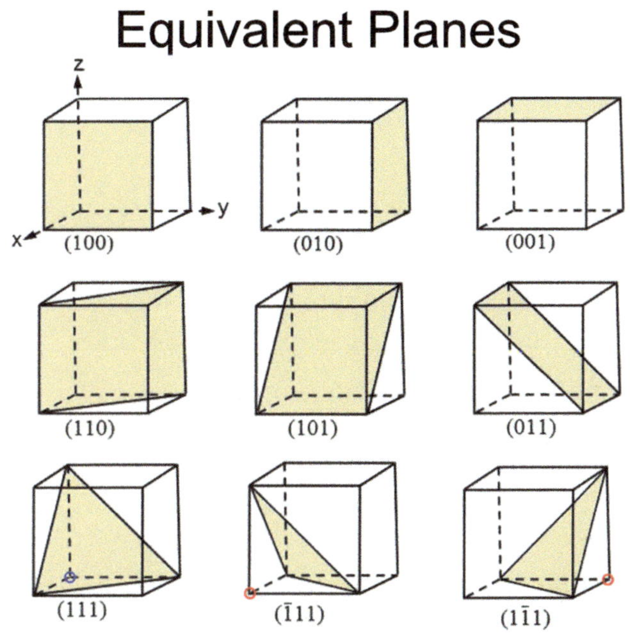

Note the shift of origin from blue to red circle for the negative indices

Planes in Hexagonal System

In the cubic system all the faces of the cube are equivalent, that is, they have similar indices.

However, this is not the case in the hexagonal system. The six prism faces for example have indices (1 0 0), (0 1 0),($\bar{1}$ 1 0), ($\bar{1}$ 0 0), (0 $\bar{1}$ 0), (1 $\bar{1}$ 0), which are not same.

In order to address this, a fourth axis (a_3) which is opposite to the vector sum of a_1 and a_2 is used and a corresponding fourth index *i* is used along with *hkl*. Therefore the indices of a plane is given by *(hkil)* whre $i = -(h+k)$. Sometime *i* is replaced with a dot and written as (h k . l)

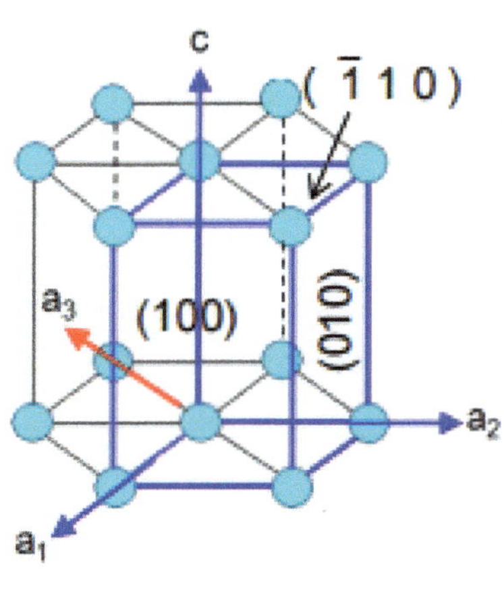

The indices of six faces now become $(1\,0\,\bar{1}\,0)$, $(0\,1\,\bar{1}\,0)$ $(\bar{1}\,1\,0\,0)$, $(\bar{1}\,0\,1\,0)$, $(0\,\bar{1}\,1\,0)$, $(1\,\bar{1}\,0\,0)$ which are now equivalent and belong to the $\{1\,0\,\bar{1}\,0\}$ family of planes.

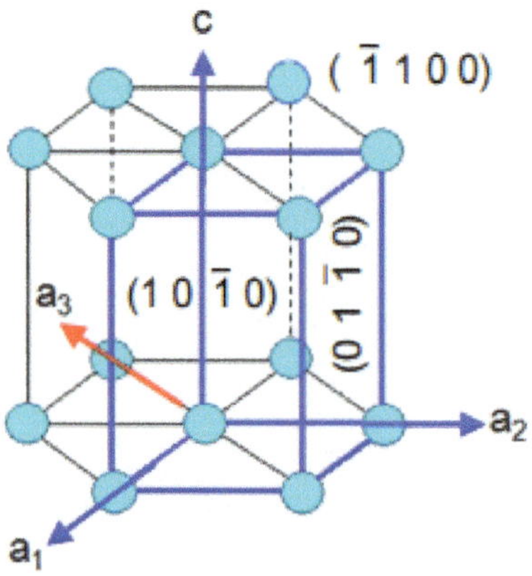

Interplanar Spacing

The spacing between planes in a crystal is known as interplanar spacing and is denoted as d_{hkl}.

In the cubic system spacing between the (hkl) planes is given as

$$\frac{1}{d^2} = \frac{1}{a^2}(h^2 + k^2 + l^2)$$

For example, d_{hkl} of {111} planes d_{111} $a/\sqrt{3}$

In Tetragonal system $\frac{1}{d_{hkl}^2} = \frac{1}{a^2}(h^2 + k^2) + \frac{1}{c^2}l^2$

In Orthorhombic system $\frac{1}{d_{hkl}^2} = \frac{1}{a^2}h^2 + \frac{1}{a^2}k^2 + \frac{1}{c^2}l^2$

In Hexagonal system $\frac{1}{d_{hkl}^2} = \frac{4}{3a^2}(h^2 + hk + k^2) + \frac{1}{c^2}l^2$

Crystal Directions

The directions in a crystal are given by specifying the coordinates (u, v, w) of a point on a vector (r_{uvw}) passing through the origin. r_{uvw} = ua + vb + wc. It is indicated as [uvw]. For example, the direction [110] lies on a vector r_{110} whose projection lengths on x and y axes are one unit (in terms of unit vectors *a* and *b*).

Directions of a form or family like [110], [101], [011] are written as <110>.

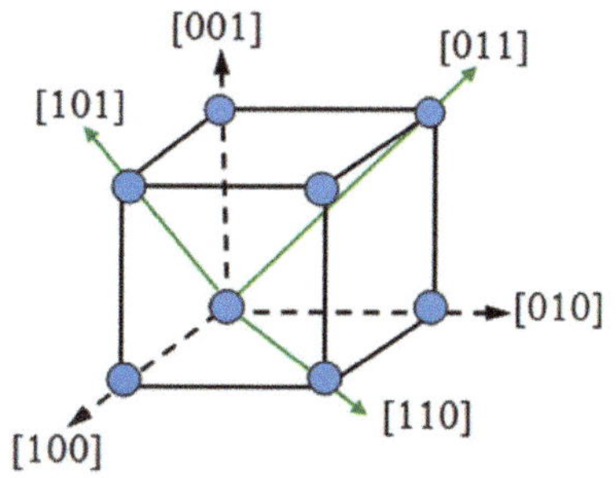

<100> and <110> family

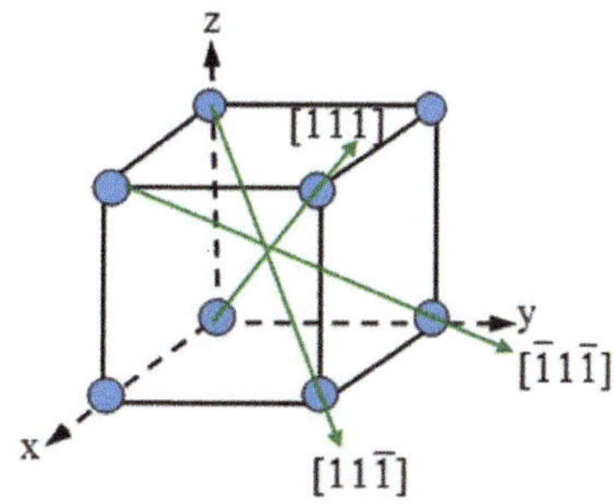

<111> family

- The line which passes through *uvw* will also pass through *2u2v2w* and ½ u ½ v ½ w. Hence [uvw], [2u2v2w] and [½ u ½ v ½ w] are same and written as [uvw].
- Fractions are converted in to integers (as shown in the figure below) and reduced to lowest terms.

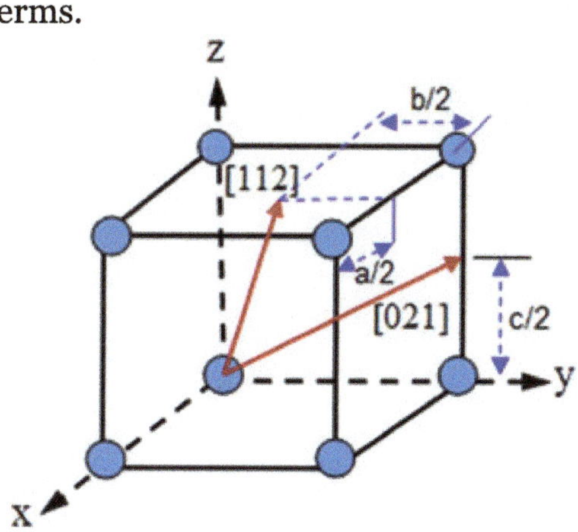

To determine a direction of a line in the crystal:

- Find the coordinates of the two ends of the line and subtract the coordinates (Head – Tail) OR draw a line from the origin parallel to the line and find its projection lengths on x, y and z axis in terms of the unit vectors a, b and c.
- Convert fractions, if any, in to integers and reduce to lowest term.
- Enclose in square brackets [uvw].

Directions in Hexagonal Crystal

Like planes, directions in the hexagonal system are also written in terms of four indices as [uvtw].

If [UVW] are indices in three axes then it can be converted to four-axis indices [uvtw] using the following relations.

$$U = u - t \qquad V = v - t \qquad W = w$$
$$u = (2U - V)/3 \qquad v = (2V - U)/3 \qquad t = -(u + v) = -(U + V)/3$$
$$w = W$$

Ex: = $[2\ \bar{1}\ \bar{1}\ 0]$, [210] = $[1\ 0\ \bar{1}\ 0]$

Relationship between Direction and Planes

In the cubic system planes and directions having same indices are perpendicular to each other i.e. if [uvw] direction is perpendicular to (hkl) plane then h = u, k = v and l = w Ex: {100} planes and <100> directions are perpendicular to each other.

If [uvw] direction is parallel to (hkl), that is if [uvw] lies in the plane (hkl) then hu + kv + lw = 0. For example, $[\ \bar{1}\ 1\ 0\ ]$ lies in the plane (111) since 1.(-1) + 1.1 + 1.0 = 0

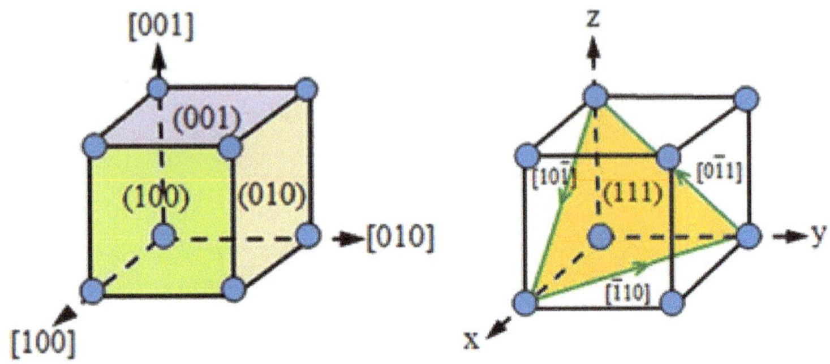

Coordination Number

Coordination number is the number of nearest neighbor to a particular atom in the crystal.

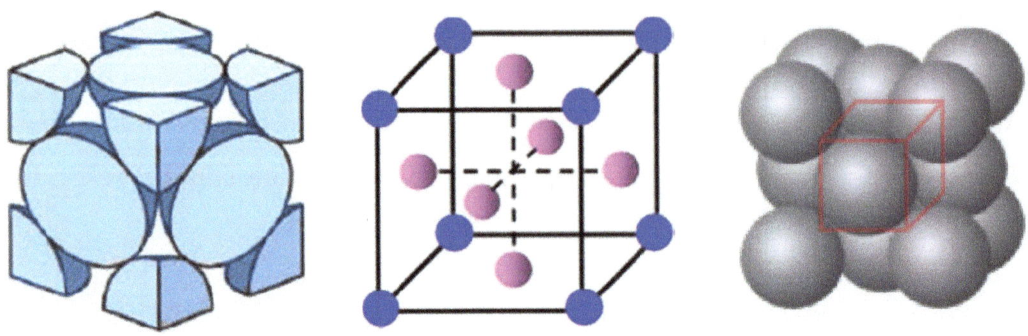

In the FCC lattice each atom is in contact with 12 neighbor atoms. FCC coordination number Z = 12

For example, the face centered atom in the front face is in contact with four corner atoms and four other face-centered atoms behind it (two sides, top and bottom) and is also touching four face-centered atoms of the unit cell in front of it.

The coordination number of BCC crystal is 8.

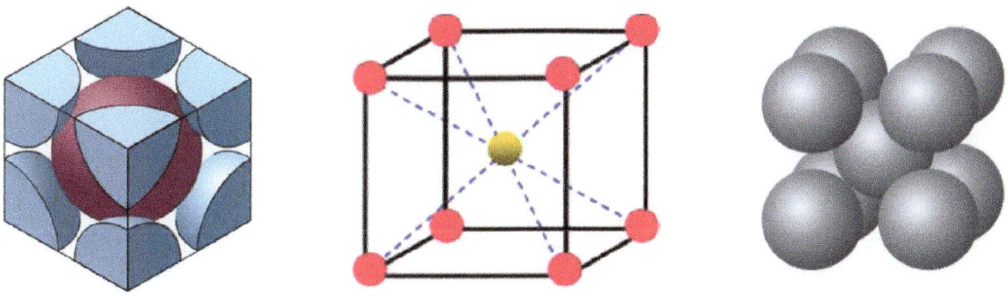

The body centered atom is in contact with all the eight corner atoms. Each corner atom is shared by eight unit cells and hence, each of these atoms is in touch with eight body centered atoms.

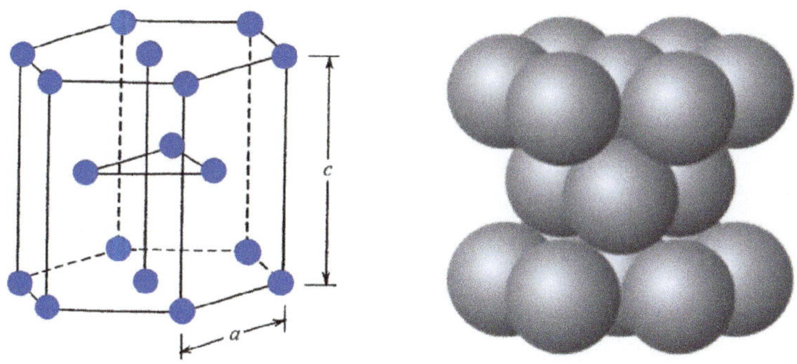

In Hexagonal lattice Z = 12. The center atom of the top face is in touch with six corner atoms, three atoms of the mid layer and other three atoms of the mid layer of the unit cell above it.

Atomic Packing Factor

Atomic packing factor (APF) or packing efficiency indicates how closely atoms are packed in a unit cell and is given by the ratio of volume of atoms in the unit cell and volume of the unit cell

$$APF = \frac{Volume\,of\,atoms}{Volume\,of\,unit\,cell}$$

FCC Lattice

In the FCC unit cell effective number of atoms = 8 corner atoms x (1/8) (each atom is shared by 8 unit cells) + 6 face- centered atoms x1/2 (each shared by two unit cells) = 4

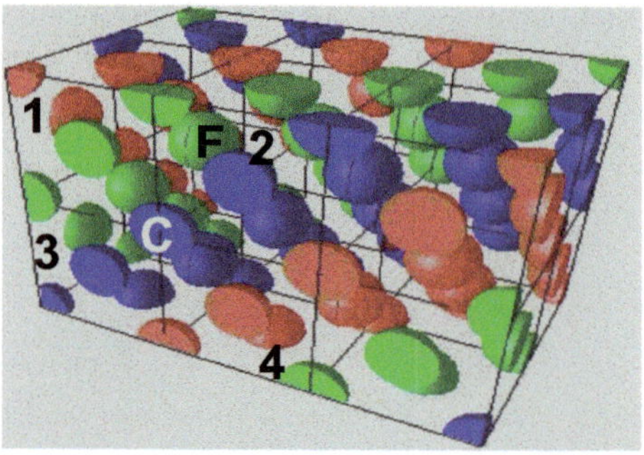

The corner atom *C* is shared by unit cells *1, 2, 3, 4* and four more in front of each of them. The face-centered atom, *F* is shared between cells *1* and *2*.

Considering the atoms as hard spheres of radius *R*

$$\text{Total volume of atoms} = 4 \times \frac{4}{3}\pi R^3$$

The relation between *R* and the FCC cell side a as shown in the figure below is $\sqrt{2}a = 4R$

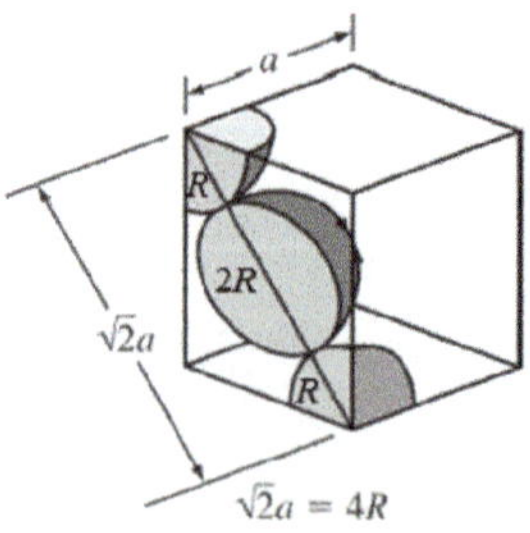

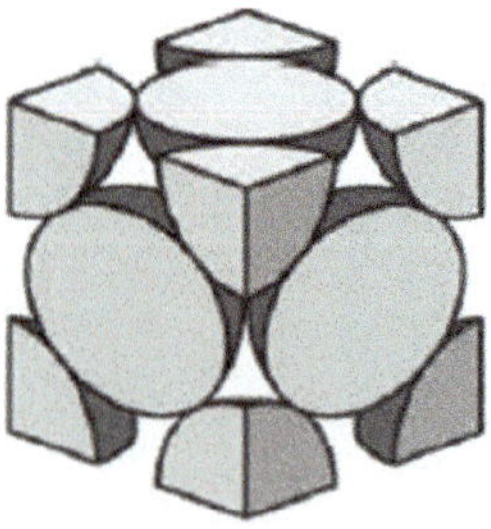

$$APF(FCC) = \frac{4 \times \frac{4}{3}\pi R^3}{a^3} = \frac{16 \times 2\sqrt{2}\pi a^3}{3 \times 64 a^3} = 0.74$$

BCC

For BCC crystals effective number of atoms per unit cell is 8 x 1/8 + 1 = 2 and the relation between *R* and *a* is $\sqrt{3}\text{a} = 4\text{R}$

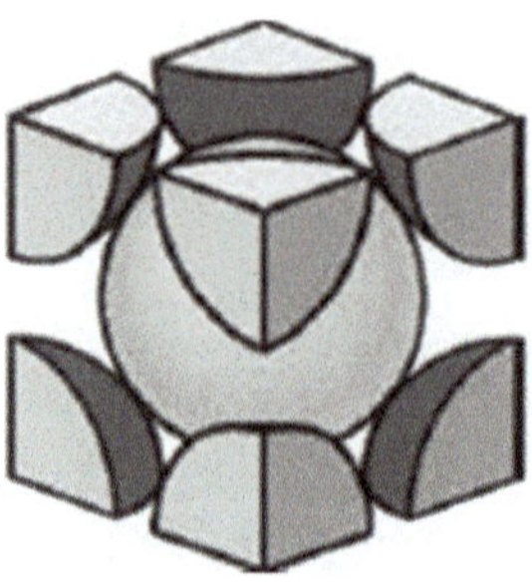

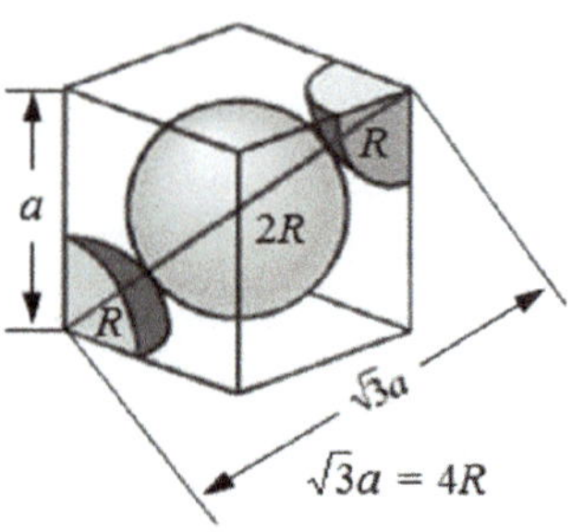

$$APF(BCC) = \frac{2 \times \frac{4}{3}\pi R^3}{a^3} = \frac{8 \times 3\sqrt{3}\pi a^3}{3 \times 64 a^3} = 0.68$$

Hexagonal Lattice

In the Hexagonal unit cell, number of atoms = 12 corner atoms × 1/6 (shared by six unit cells) + Two face atoms × ½ + 3 interior = 6.

$$2R = a$$

Unit cell volume = (6 × ½ × a × h) × c = (3 × a × a sin60°) × c = $3a^2c$ sin60°

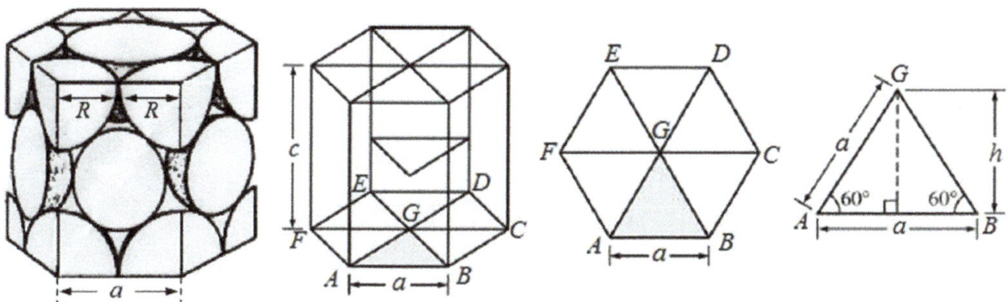

Atomic Packing Factor

The face-centered atom and the three mid-layer atoms form a tetrahedron MNOP which has sides equal to *a* (as atoms at vertices touch each other) and height of c/2. Using this tetrahedron it can be shown that for an ideal hexagonal crystal c/a ratio = 1.633

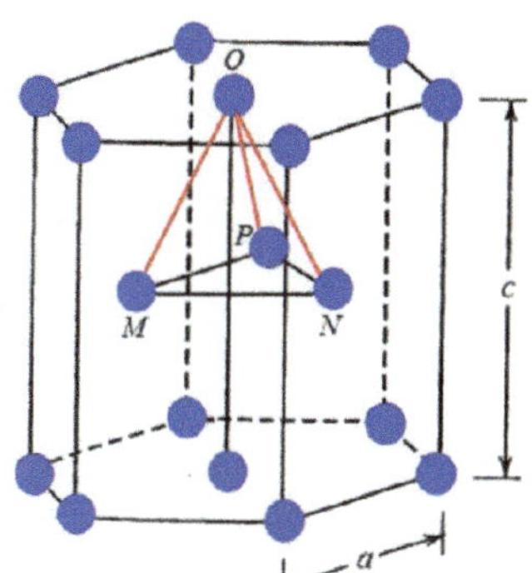

$$APF(HCP) = \frac{6 \times \frac{4}{3}\pi R^3}{3a^2 c \sin 60^{\circ}} = \frac{8\pi a^3}{3 \times 8 \times 1.414 a^3} = 0.74$$

Planar Density

Planar density (PD) refers to density of atomic packing on a particular plane.

$$Planar\ Density = \frac{Number\ of\ atoms\ on\ a\ plane}{Area\ of\ plane}$$

For example, there are 2 atoms (1/4 × 4 corner atoms + 1/2 × 2 side atoms) in the {110} planes in the FCC lattice. Planar density of {110} planes in the FCC crystal

$$PD_{(110)} = \frac{2}{a\sqrt{2}a} = \frac{\sqrt{2}}{a^2}$$

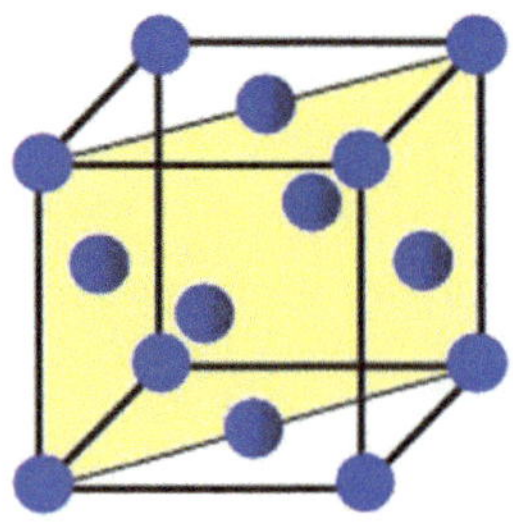

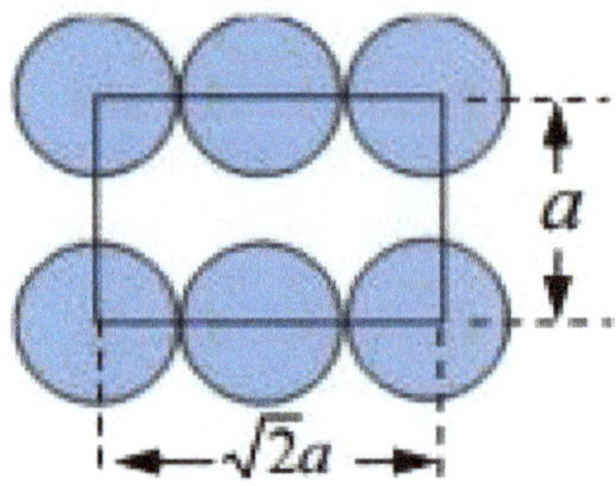

In the {111} planes of the FCC lattice there are 2 atoms (1/6 × 3 corner atoms + 1/2 × 3 side atoms). Planar density of {111} planes in the FCC crystal

$$PD_{(111)} = \frac{2}{\frac{1}{2}\sqrt{2}a \times \sqrt{2}a\frac{\sqrt{3}}{2}} = \frac{4}{\sqrt{3}a^2}$$

This is higher than {110} and any other plane. Therefore, {111} planes are most densely packed planes in the FCC crystal

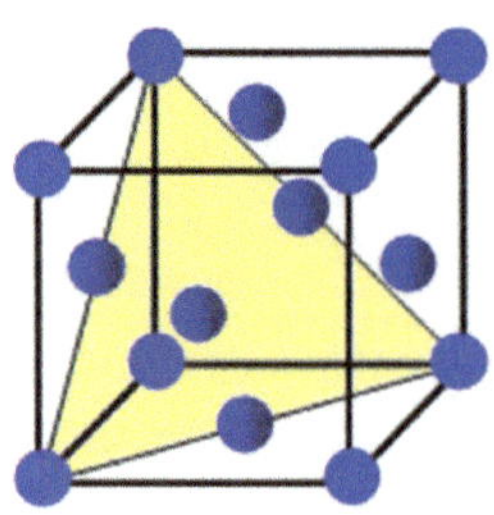

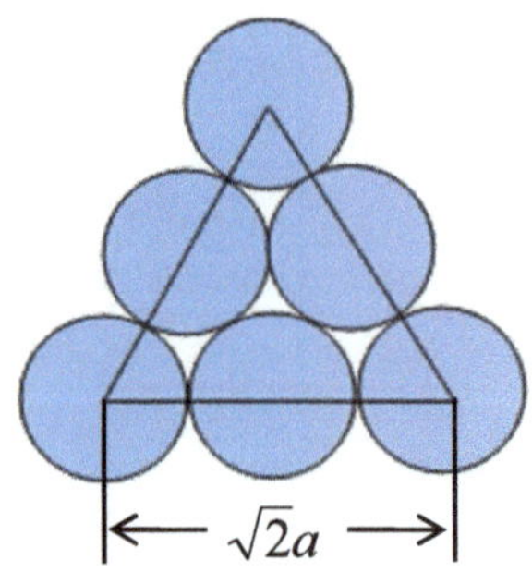

Linear Density

Linear density (LD) is the number of atoms per unit length along a particular direction

$$LinearDensity = \frac{Number\ of\ atoms\ on\ the\ direction\ vector}{Length\ of\ the\ direction\ vector}$$

<110> directions in the FCC lattice have 2 atoms (1/2 x 2 corner atoms + 1 center atom) and the length is $\sqrt{2}a$

$$LD_{[110]} = \frac{2}{\sqrt{2}a} = \frac{\sqrt{2}}{a}$$

This is the most densely packed direction in the FCC lattice

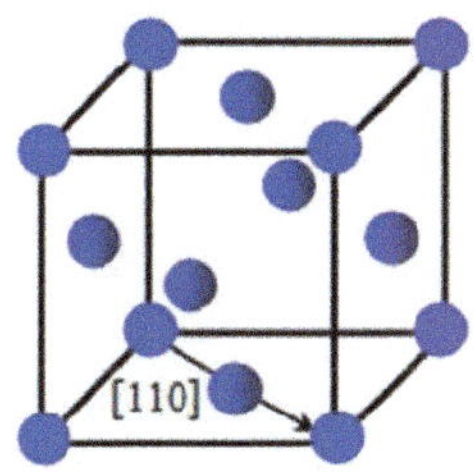

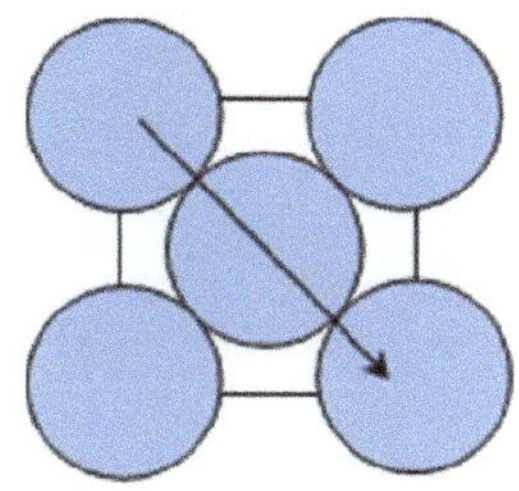

Close-Packed Structure

- FCC and hexagonal crystal structures are most highly packed with packing efficiency of 74% (APF= 0.74). Such structures can be described in terms of close-packed atomic planes.
- In FCC, {111} planes are close-packed and the basal plane (0001) is the close-packed one in hexagonal close-packed (HCP) system. Therefore, both of these structures can be generated by stacking of these planes. A portion of such a stack is shown in the picture below.

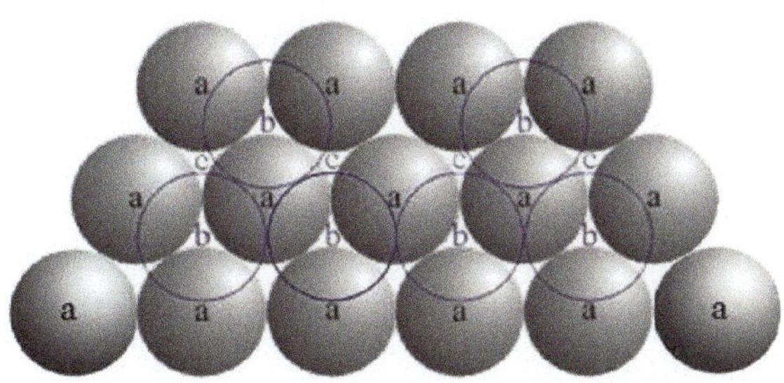

- There are two types of voids between the atoms – vertex up (*b*), and vertex down (*c*). The atoms in the next layer sit on the *b* sites.
- In FCC, atoms in the third layer sit over the *c* sites and this repeats giving rise to ABCABCABC type of stacking.

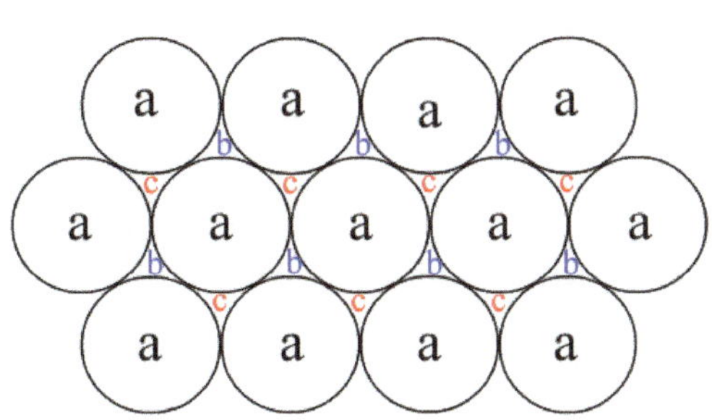

- In HCP system, centers of atoms of the third layer lie directly over centers of atoms of the first layer (a positions) giving rise to ABABAB type of stacking.

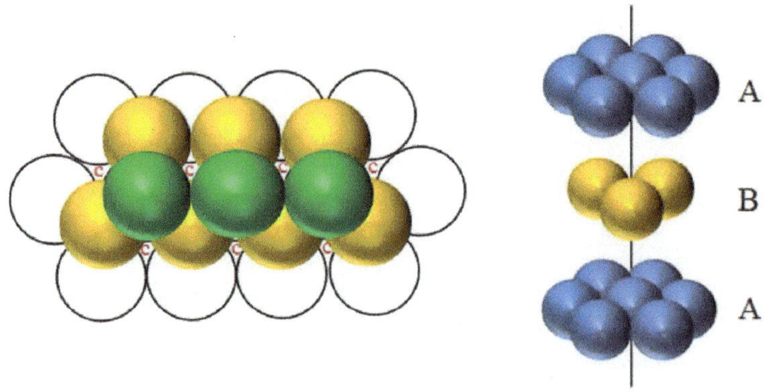

HCP

Structure-Property Correlation

Aluminium (Al) is ductile while iron (Fe) and magnesium (Mg) are not. This can be explained from their crystal structures.

Al is FCC where as Fe is BCC and Mg is HCP.

Plastic deformation in metals takes place mainly by a process called slip. Slip can broadly be visualized as sliding of crystal planes over one another. Slip occurs on most densely packed planes in the most closely packed directions lying on that plane.

The slip plane and the direction together is called a Slip system

- In FCC, {111} planes are close-packed and there are four unique {111} planes. Each of these planes contains three closely packed <110> directions. Therefore, there are 4 x 3 = 12 slip systems
- In HCP, the basal plane, (0001) is the close-packed and it contains three <1 1 $\bar{2}$ 0> directions. Hence, number of slip system = 1 x 3 = 3
- Slip in more number of slip systems allows greater plastic deformation before fracture imparting ductility to FCC materials

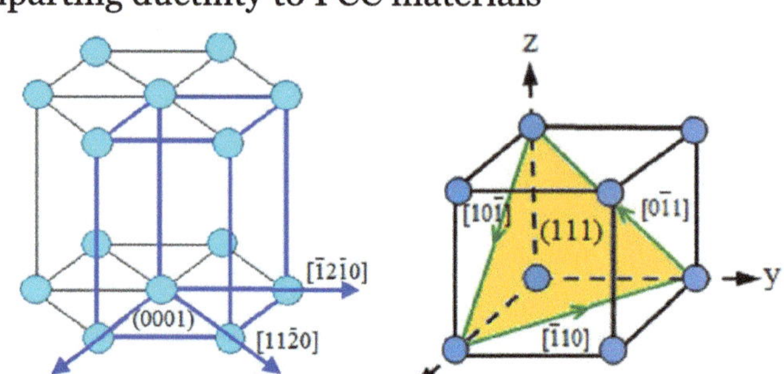

- Close-packed planes are also planes with greatest interplanar spacing and this allows slip to take place easily on these planes.
- BCC structure on the other hand has 48 possible slip systems. However, there is no close-packed plane. Hence, plastic deformation before fracture is not significant. Slip might occur in {110}, {112} and {123} planes in the <111> directions.

Polycrystalline

- Polycrystalline material – Aggregate of several crystals or grains
- The boundary between the grain is the grain boundary across which the orientation of the crystal changes.
- The point at which three boundaries meet is called the triple junction.

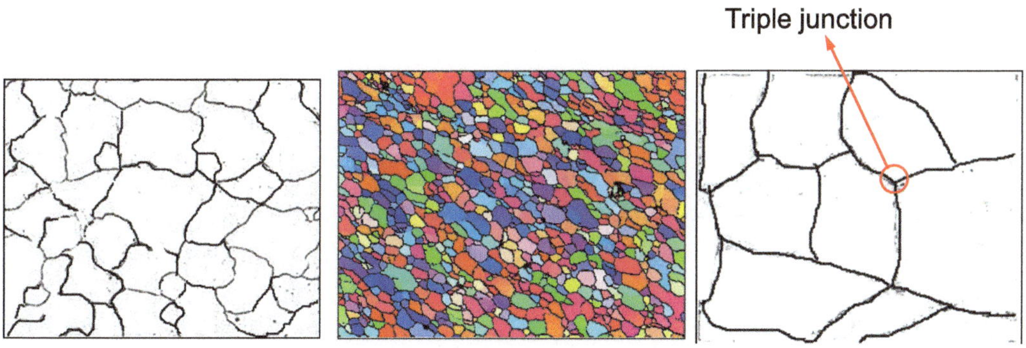

Grain Size Determination

Line intercept method

This is one of the most commonly used methods. Number of grains intersecting a given length of a random line is counted.

Grain size D = Length of the line/no of grains intersected

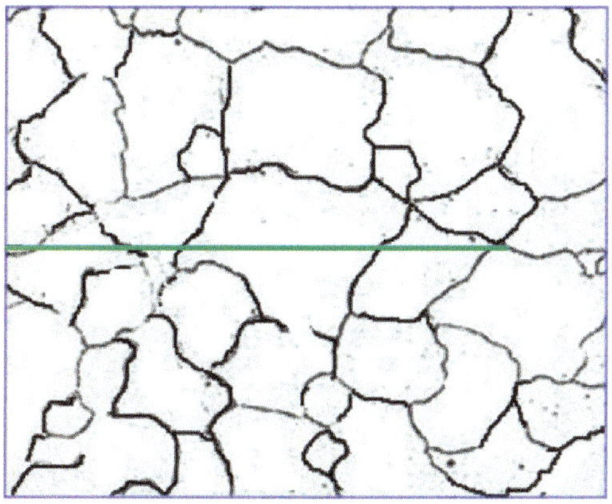

Grain Size Measurement

ASTM grain size number, G – Number of grains per unit area at a particular magnification

G = -2.9542 + 1.4427 ln N where, N is number of grains/mm²

G is compared in ASTM grain size chart to obtain the grain size.

ASTM No.	Grains/mm²	Grains/mm³	Avg. grain size, mm
-1	3.9	6.1	0.51
0	7.8	17.3	0.36
1	15.5	49.0	0.25
2	31.0	138	0.18
3	62.0	391	0.125
4	124	1105	0.09
5	248	3126	0.065
6	496	8842	0.045
7	992	25010	0.032
8	1980	70700	0.022
9	3970	200000	0.016
10	7940	566000	0.011

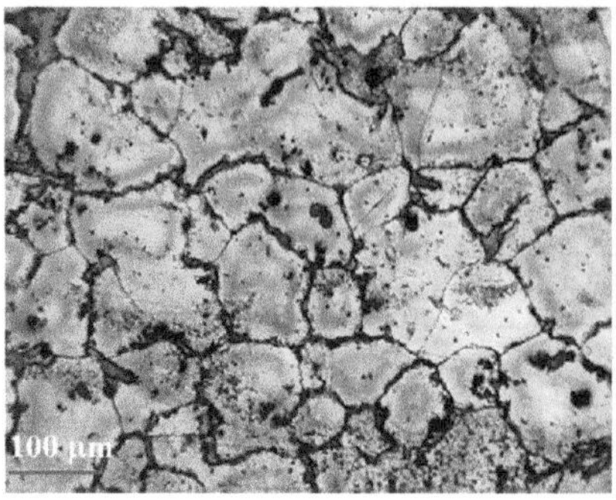

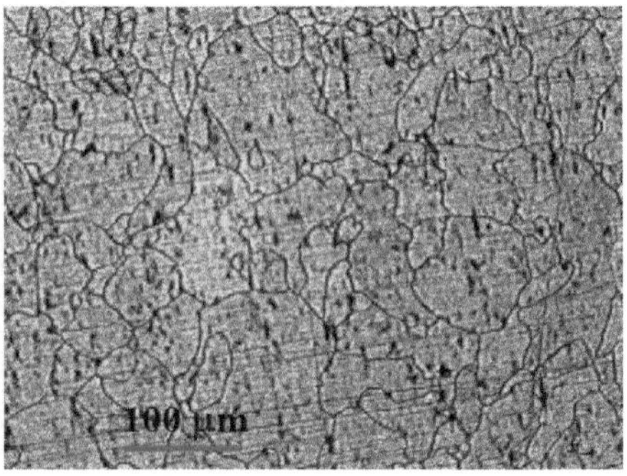

Calculate the grain size from the micrographs using the line intercept method

Grain size – Property Relationship

Hall-Petch Relation

A general relationship between mechanical properties and grain size is given by the Hall-Petch equation

$$\sigma_0 = \sigma_i = kd^{-1/2}$$

σ_o is the yield strength, d is the grain size and σ_i and k are material dependent constants.

Finer grain size means more grain boundaries or higher grain boundary area per unit volume. Deformation in metals takes place by dislocation motion and grain boundaries act as obstacles to dislocation motion. Hence, presence of more grain boundaries (finer grain size) will increase the resistance to deformation and enhance the strength.

Single Crystals

- Single crystal – only one grain or crystal and hence, no grain boundaries.
- Useful for applications where grain boundaries are harmful. For example, high temperature deformation or creep resistance (as creep takes place by grain boundary sliding) - single crystal turbine blades

Non Crystalline or Amorphous Materials

Amorphous – random arrangement of atoms: Silicate glass, Polymers

Metallic Amorphous Materials

Bulk Metallic Glass (BMG)

A metallic system can be made amorphous by decreasing the chance of crystallization: – Allow less time for crystallization during solidification – Rapid solidification processing (RSP)

Increase confusion for a particular crystal form by increasing the number of components (alloying elements).

Multi component – elements with different chemical nature and atomic size and number together reduce the chance of crystallization

$Zr_{41.2}Ti_{13.8}Cu_{12.5}Ni_{10}Be_{22.5}[=(Zr_3Ti)_{55}(Be_9Cu_5Ni_4)_{45}]$, commonly referred to as Vitreloy 1 (Vit1)

Processing routes:

RSP - Melt spinning

Ball milling

Solid-state amorphisation reaction

Nuclear irradiation

Properties of BMGs

Mechanical properties – High strength, low ductility

Formation of small crystallites in glassy matrix by annealing can improve ductility.

Some Properties of Vit1

Elastic strain	Tensile strength	Young's Modulus	Shear Modulus	Hardness	Fracture toughness
2%	1.9 GPa	96 GPa	34.3 GPa	534 VHN	55 MPa $m^{1/2}$

Magnetic properties – excellent soft magnetic properties due to absence of crystalline magnetic anisotropy FINEMET - Fe-Si-B-Cu-Nb

Excellent corrosion resistance

Good acoustic properties

Applications of BMGs

Golf heads – BMG golf heads can transfer 99% energy compared to 60% for steel heads and 70% for Ti heads

Thin yet strong electronic casing – MP3 player, mobile phones, digital cameras, PDA

Surgical instruments, Prosthetic implants (Biomaterials)

Sensors for electronic article surveillance, cores for high frequency transformers

Self-sharpening tank armor penetrator

Nano Materials

Nanometer = 10^{-9}m

micrometer = 10^{-6}m

mm = 10^{-3} m

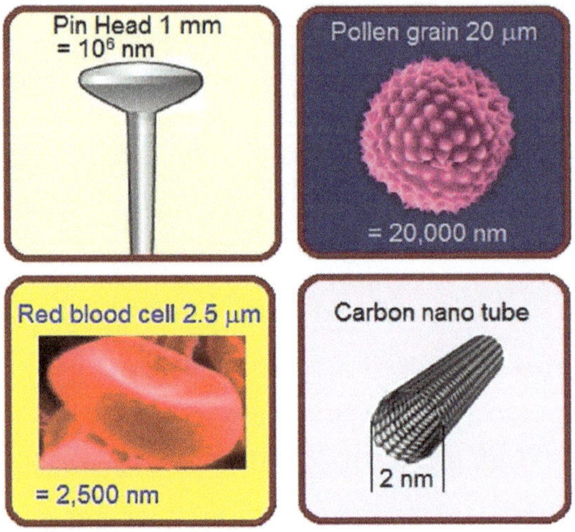

- Nano Structured Material (NSM) – At least one component of the microstructure e.g. grains, particles or dispersoids, is nano meter in size.
- Nano particles possess very high surface area /unit volume which gives rise to unique physical and chemical properties.
- In a nano-grained material the grain boundary width is comparable with the grain size.

Nano particles

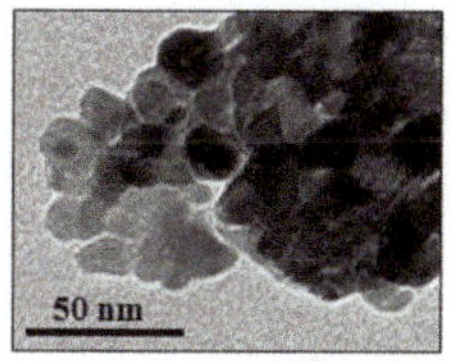

Nano Ni-ZrO_2 composite

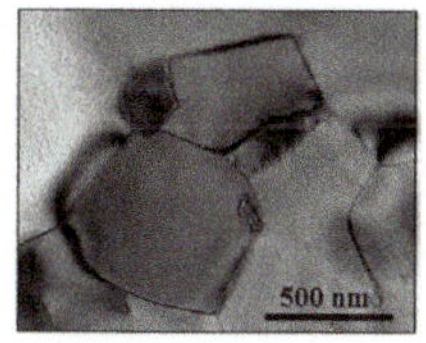

Nano grains in Al

Nano -Structured Materials (NSM)

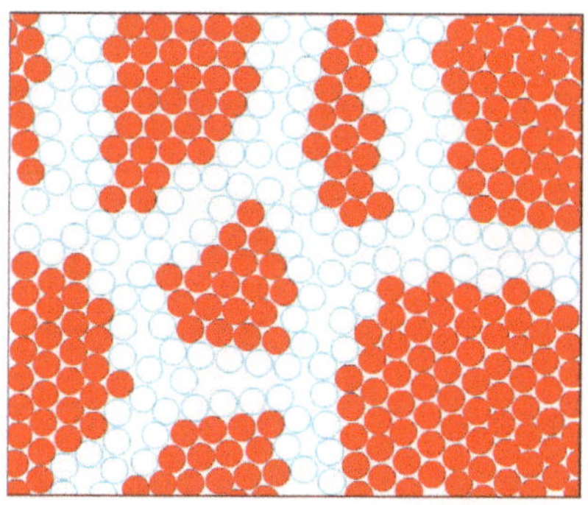

A two-dimensional representation of a nanostructured material. Red circles indicate atoms in the grain while open circles indicate atoms in the grain boundary region

Classification of NSMs

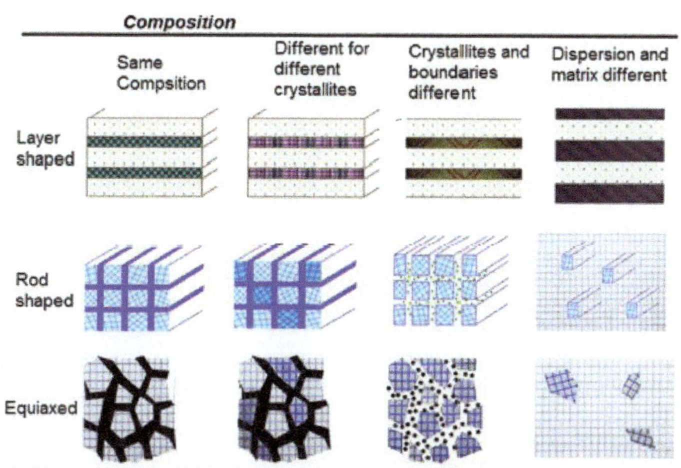

Classification of NSMs according to their chemical composition and shape

Processing Routes

- Inert gas condensation
- Vapour phase condensation
- Wet chemical method
- High energy mechanical alloying
- Plasma processing
- Combustion synthesis
- Super critical liquid
- Chemical vapor deposition (CVD)
- Physical vapor deposition (PVD)

Properties of NSMs

Superior functional properties – Photoluminescence, Electroluminescence, Electronic and magnetic peoperties

Enhanced catalytic activity – Very high surface area per unit volume

Structural properties – Very high strength – Hall-petch equation $\sigma_0 = \sigma_i = kd^{-1/2}$

Ultrafine grain size - Superplasticity

High Hardness – Nanocrystalline WC-Co composites High stiffness - CNTs

Applications

Sunscreens Lotions - Many sunscreens contain nano particles of zinc oxide or titanium oxide.

Self-cleaning glass: Activ Glass - uses nanoparticles to make the glass photocatalytic and hydrophilic.

- Clothing: coating fabrics with a thin layer of zinc oxide nanoparticles, better protection from UV radiation.
- Stain resistant clothes - have nanoparticles in the form of little hairs or whiskers that help repel water and other materials

Glossy colors – Nano particles of pigments ($CoAl_2O_4$). Shiny, better looking colors for cars

Scratch resistant coatings – addition of aluminum silicate nanoparticles to scratch-resistant polymer coatings. Scratch- resistant coatings for cars and eye lenses.

NSMs in Electronics – Better and Smaller

- The advent of the nano technology has immensely helped in miniaturization which is an essential part in many electronic gadgets.
- The nano technology is already in use in many electronic gadgets like mobile phones, I-pads, cameras, palm tops and so on.
- Quantum dots - nano-scale semiconductor crystals. Dramatic improvement in digital storage (Flash memory).
- Less power consumption, better resolution (in visuals), smaller size.

Applications of NSMs

- Flash memory: Conventional – tunneling film must be sufficiently insulating to store the charge. Limits thickness reduction and thus miniaturization. Cannot work with defects in the film.
- Nano dot or quantum dot flash memory allows for higher defect tolerance, lower thickness (and hence miniaturization). Less power consumption

Quantum dots display -By altering the size of the nano particles, the color they emit can be changed.

6 nm – red, 2 nm - blue

Quantum dot display in place of LCDs or OLEDs – less power consumption, better quality pictures.

Crystallographic Defect

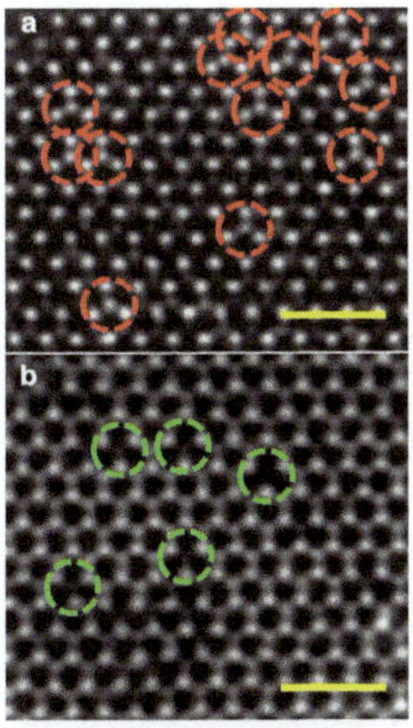

Electron microscopy of antisites (a, Mo substitutes for S) and vacancies (b, missing S atoms) in a monolayer of molybdenum disulfide. Scale bar: 1 nm.

Crystalline solids exhibit a periodic crystal structure. The positions of atoms or molecules occur on repeating fixed distances, determined by the unit cell parameters. However, the arrangement of atoms or molecules in most crystalline materials is not perfect. The regular patterns are interrupted by crystallographic defects.

Point Defects

Point defects are defects that occur only at or around a single lattice point. They are not extended in space in any dimension. Strict limits for how small a point defect is are generally not defined explicitly. However, these defects typically involve at most a few extra or missing atoms. Larger defects in an ordered structure are usually considered dislocation loops. For historical reasons, many point defects, especially in ionic crystals, are called *centers*: for example a vacancy in many ionic solids is called a luminescence center, a color center, or F-center. These dislocations permit ionic transport through crystals leading to electrochemical reactions. These are frequently specified using Kröger–Vink Notation.

- Vacancy defects are lattice sites which would be occupied in a perfect crystal, but are vacant. If a neighboring atom moves to occupy the vacant site, the vacancy moves in the opposite direction to the site which used to be occupied by the moving atom. The stability of the surrounding crystal structure guarantees that the neighboring atoms will not simply collapse around the vacancy. In some materials, neighboring atoms actually move away from a vacancy, because they experience attraction from atoms in the surroundings. A vacancy (or pair of vacancies in an ionic solid) is sometimes called a Schottky defect.
- Interstitial defects are atoms that occupy a site in the crystal structure at which there is usually not an atom. They are generally high energy configurations. Small atoms in some crystals can occupy interstices without high energy, such as hydrogen in palladium.

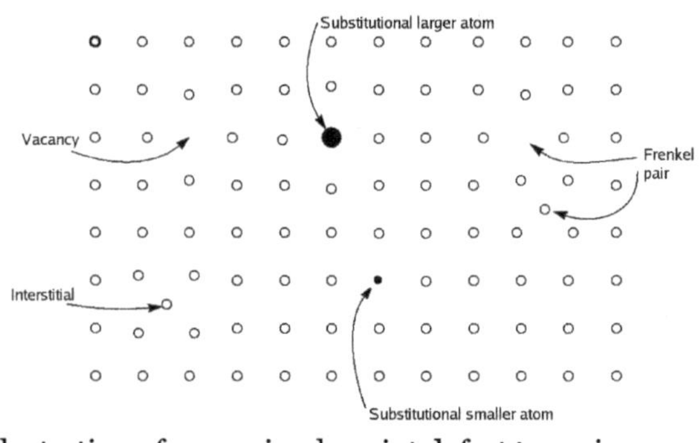

Schematic illustration of some simple point defect types in a monatomic solid

- A nearby pair of a vacancy and an interstitial is often called a Frenkel defect or Frenkel pair. This is caused when an ion moves into an interstitial site and creates a vacancy.

- Due to fundamental limitations of material purification methods, materials are never 100% pure, which by definition induces defects in crystal structure. In the case of an impurity, the atom is often incorporated at a regular atomic site in the crystal structure. This is neither a vacant site nor is the atom on an interstitial site and it is called a *substitutional* defect. The atom is not supposed to be anywhere in the crystal, and is thus an impurity. In some cases where the radius of the substitutional atom (ion) is substantially smaller than that of the atom (ion) it is replacing, its equilibrium position can be shifted away from the lattice site. These types of substitutional defects are often referred to as off-center ions. There are two different types of substitutional defects: Isovalent substitution and aliovalent substitution. Isovalent substitution is where the ion that is substituting the original ion is of the same oxidation state as the ion it is replacing. Aliovalent substitution is where the ion that is substituting the original ion is of a different oxidation state than the ion it is replacing. Aliovalent substitutions change the overall charge within the ionic compound, but the ionic compound must be neutral. Therefore, a charge compensation mechanism is required. Hence either one of the metals is partially or fully oxidised or reduced, or ion vacancies are created.

- Antisite defects occur in an ordered alloy or compound when atoms of different type exchange positions. For example, some alloys have a regular structure in which every other atom is a different species; for illustration assume that type A atoms sit on the corners of a cubic lattice, and type B atoms sit in the center of the cubes. If one cube has an A atom at its center, the atom is on a site usually occupied by a B atom, and is thus an antisite defect. This is neither a vacancy nor an interstitial, nor an impurity.

- Topological defects are regions in a crystal where the normal chemical bonding environment is topologically different from the surroundings. For instance, in a perfect sheet of graphite (graphene) all atoms are in rings containing six atoms. If the sheet contains regions where the number of atoms in a ring is different from six, while the total number of atoms remains the same, a topological defect has formed. An example is the Stone Wales defect in nanotubes, which consists of two adjacent 5-membered and two 7-membered atom rings.

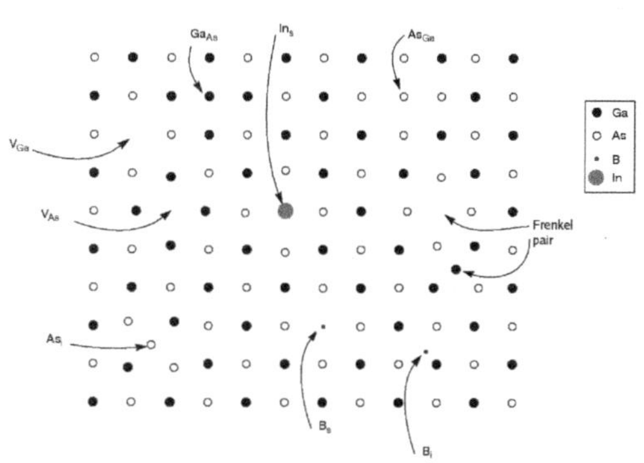

Schematic illustration of defects in a compound solid, using GaAs as an example.

- Also amorphous solids may contain defects. These are naturally somewhat hard to define, but sometimes their nature can be quite easily understood. For instance, in ideally bonded amorphous silica all Si atoms have 4 bonds to O atoms and all O atoms have 2 bonds to Si atom. Thus e.g. an O atom with only one Si bond (a dangling bond) can be considered a defect in silica. Moreover, defects can also be defined in amorphous solids based on empty or densely packed local atomic neighbourhoods, and the properties of such 'defects' can be shown to be similar to normal vacancies and interstitials in crystals,.

- Complexes can form between different kinds of point defects. For example, if a vacancy encounters an impurity, the two may bind together if the impurity is too large for the lattice. Interstitials can form 'split interstitial' or 'dumbbell' structures where two atoms effectively share an atomic site, resulting in neither atom actually occupying the site.

Line Defects

Line defects can be described by gauge theories.

Dislocations are linear defects, around which (the dislocation line) some of the atoms of the crystal lattice are misaligned. There are two basic types of dislocations, the *edge* dislocation and the *screw* dislocation. "Mixed" dislocations, combining aspects of both types, are also common.

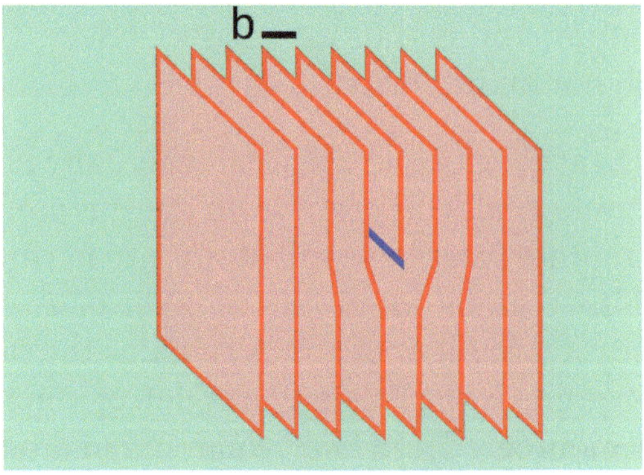

An *edge dislocation* is shown. The dislocation line is presented in blue, the Burgers vector b in black.

Edge dislocations are caused by the termination of a plane of atoms in the middle of a crystal. In such a case, the adjacent planes are not straight, but instead bend around the edge of the terminating plane so that the crystal structure is perfectly ordered on either side. The analogy with a stack of paper is apt: if a half a piece of paper is inserted in a stack of paper, the defect in the stack is only noticeable at the edge of the half sheet.

The screw dislocation is more difficult to visualise, but basically comprises a structure in which a helical path is traced around the linear defect (dislocation line) by the atomic planes of atoms in the crystal lattice.

The presence of dislocation results in lattice strain (distortion). The direction and magnitude of such distortion is expressed in terms of a Burgers vector (b). For an edge type, b is perpendicular to the dislocation line, whereas in the cases of the screw type it is parallel. In metallic materials, b is aligned with close-packed crystallographic directions and its magnitude is equivalent to one interatomic spacing.

Dislocations can move if the atoms from one of the surrounding planes break their bonds and rebond with the atoms at the terminating edge.

It is the presence of dislocations and their ability to readily move (and interact) under the influence of stresses induced by external loads that leads to the characteristic malleability of metallic materials.

Dislocations can be observed using transmission electron microscopy, field ion microscopy and atom probe techniques. Deep level transient spectroscopy has been used for studying the electrical activity of dislocations in semiconductors, mainly silicon.

Disclinations are line defects corresponding to "adding" or "subtracting" an angle around a line. Basically, this means that if you track the crystal orientation around the line defect, you get a rotation. Usually, they were thought to play a role only in liquid crystals, but recent developments suggest that they might have a role also in solid materials, e.g. leading to the self-healing of cracks.

Planar Defects

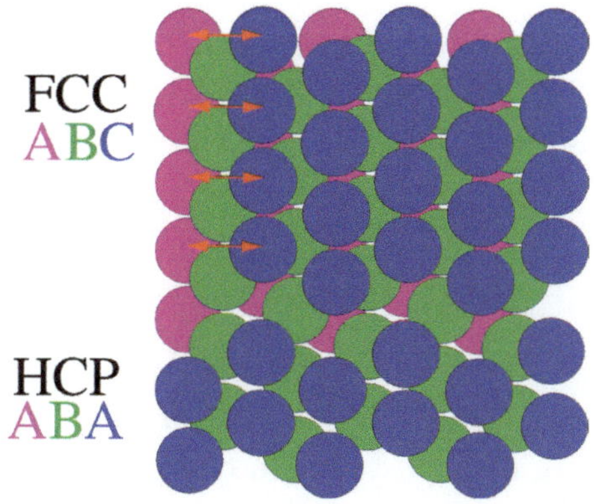

Origin of stacking faults: Different stacking sequences of close-packed crystals

- Grain boundaries occur where the crystallographic direction of the lattice abruptly changes. This usually occurs when two crystals begin growing separately and then meet.
- Antiphase boundaries occur in ordered alloys: in this case, the crystallographic direction remains the same, but each side of the boundary has an opposite phase: For example, if the ordering is usually ABABABAB (hexagonal close-packed crystal), an antiphase boundary takes the form of ABABBABA.

- Stacking faults occur in a number of crystal structures, but the common example is in close-packed structures. They are formed by a local deviation of the stacking sequence of layers in a crystal. An example would be the ABABCABAB stacking sequence.
- A twin boundary is a defect that introduces a plane of mirror symmetry in the ordering of a crystal. For example, in cubic close-packed crystals, the stacking sequence of a twin boundary would be ABCABCBACBA.
- On surfaces of single crystals, steps between atomically flat terraces can also be regarded as planar defects. It has been shown that such defects and their geometry have significant influence on the adsorption of organic molecules

Bulk Defects

- Three-dimensional macroscopic or bulk defects, such as pores, cracks, or inclusions
- Voids — small regions where there are no atoms, and which can be thought of as clusters of vacancies
- Impurities can cluster together to form small regions of a different phase. These are often called precipitates.

Mathematical Classification Methods

A successful mathematical classification method for physical lattice defects, which works not only with the theory of dislocations and other defects in crystals but also, e.g., for disclinations in liquid crystals and for excitations in superfluid ^{3}He, is the topological homotopy theory.

Computer Simulation Methods

Density functional theory, classical molecular dynamics and kinetic Monte Carlo simulations are widely used to study the properties of defects in solids with computer simulations. Simulating jamming of hard spheres of different sizes and/or in containers with non-commeasurable sizes using the Lubachevsky–Stillinger algorithm can be an effective technique for demonstrating some types of crystallographic defects.

Solid Solution

- When two metals are mixed together they form an alloy if one metal is soluble in the other one in solid state. Therefore, an alloy is a solid solution of two or more metals.

- Primarily there are two types of solid solutions -
  - o Substitutional – Solute atoms occupy the regular lattice sites of the parent metal (solvent). Substitutional solid solutions can be random (Cu-Ni) or ordered (Cu-Au).
  - o Interstitial – Solute atoms occupy the interstitial positions (Steel – C solute atoms in Fe).

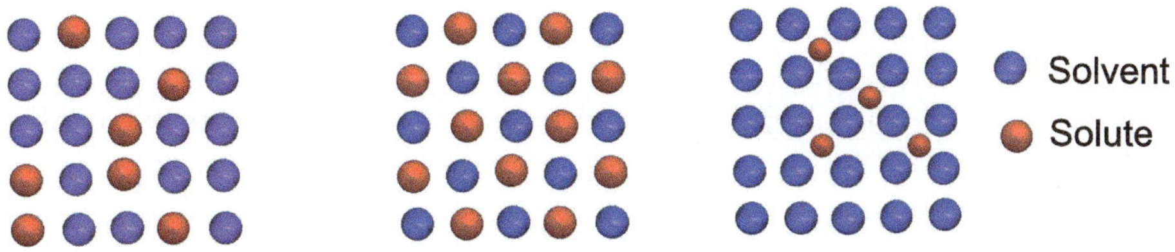

Hume-Rothery Rules

Formation of substitutional solid solutions between two metals is governed by a set of rules known as Hume-Rothery rules

- Size difference between the atoms of solute and the parent metal should be less than 15%.
- The electronegetivity difference between the metals should be small (minimum chemical affinity to each other).
- The solubility of a metal with higher valence in a solvent of lower valence is more compared to the reverse situation e.g. Zn is much more soluble in Cu than Cu in Zn.
- For complete solubility over the entire range of compositions the crystal structures of the solute and the solvent must be the same.

Ordering in Solid Solutions

- As stated before substitutional solid solutions can be either ordered or random. This depends on a thermodynamic parameter called enthalpy of mixing, ΔH_{mix}

$$\Delta G_{mix} = \Delta H_{mix} - T\Delta S_{mix}$$

  ΔG_{mix} is the Gibbs free energy change and ΔS_{mix} entropy of mixing. For an ideal solution $\Delta H_{mix} = 0$.

- If $\Delta H_{mix} > 0$, formation of like bonds (A-A or B-B) is preferred in a solid solution between metals A and B. This known as clustering.

- If $\Delta H_{mix} < 0$, unlike bonds (A-B) are preferred. This leads to ordering which may exist over short range or long range (at lower temperatures).

Intermediate Phases

- As the name suggest intermediate structures formed between two metals are neither the parent metals nor like an alloy.

Intermetallic Compounds

If two elements have high difference in electronegetivity, they tend to from a system called intermetallic compound.

Intermetallic compounds like MgSe, PbSe, Mg_2Si, Cu_2S are cubic whereas NiAs, MnSe, CuSn are hexagonal.

Electron or Hume Rothery phases

These compounds have wide range of solubility and occur at certain values of valence electrons to atom ratio such as 3:2 (CuZn), 21:13 (Cu_5Zn_8), 7:4 ($CuZn_3$).

Laves Phase

Laves phases have a general formula of AB_2, for example $MgCu_2$ (cubic), $MgZn_2$ (hexagonal), $MgNi_2$ (hexagonal).

Sigma Phase

Sigma phase has a very complex crystal structure and is very brittle. This phase can act as a source of embrittlement in some alloys such as steels.

Metal Carbides and Nitrides

Metals which have high chemical affinity for carbon and nitrogen form carbides and nitrides such as VC, NbC, VN, NbN, TiC, TiN. They can act as source of hardening in many alloys.

References

- Mermin, N. (1979). "The topological theory of defects in ordered media". Reviews of Modern Physics. 51 (3): 591–648. Bibcode:1979RvMP...51..591M. doi:10.1103/RevModPhys.51.591
- Schwartz, Mel (2002). "Tin and Alloys, Properties". Encyclopedia of Materials, Parts and Finishes (2nd ed.). CRC Press. ISBN 1-56676-661-3.
- Bernal, J. D.; Fowler, R. H. (1933). "A Theory of Water and Ionic Solution, with Particular Reference to Hydrogen and Hydroxyl Ions". The Journal of Chemical Physics. 1 (8): 515. Bibcode:-1933JChPh...1..515B. doi:10.1063/1.1749327

- Hogan, C. M. (1969). "Density of States of an Insulating Ferromagnetic Alloy". Physical Review. 188 (2): 870. Bibcode:1969PhRv..188..870H. doi:10.1103/PhysRev.188.870
- Hook, J.R.; Hall, H.E. (2010). Solid State Physics. Manchester Physics Series (2nd ed.). John Wiley & Sons. ISBN 9780471928041
- Pauling, Linus (1947). "Atomic Radii and Interatomic Distances in Metals". Journal of the American Chemical Society. 69 (3): 542. doi:10.1021/ja01195a024
- Zhang, X. Y.; Suhl, H (1985). "Spin-wave-related period doublings and chaos under transverse pumping". Physical Review A. 32 (4): 2530–2533. Bibcode:1985PhRvA..32.2530Z. PMID 9896377. doi:10.1103/PhysRevA.32.2530
- Donald E. Sands (1994). "§4-2 Screw axes and §4-3 Glide planes". Introduction to Crystallography (Reprint of WA Benjamin corrected 1975 ed.). Courier-Dover. pp. 70–71. ISBN 0486678393
- "Direct and reciprocal lattices". CSIC Dept de Cristalografia y Biologia Estructural. 6 Apr 2017. Retrieved 18 May 2017
- Stillinger, Frank H.; Lubachevsky, Boris D. (1995). "Patterns of broken symmetry in the impurity-perturbed rigid-disk crystal". Journal of Statistical Physics. 78 (3–4): 1011–1026. Bibcode:-1995JSP....78.1011S. doi:10.1007/BF02183698
- Molodets, A. M.; Nabatov, S. S. (2000). "Thermodynamic Potentials, Diagram of State, and Phase Transitions of Tin on Shock Compression". High Temperature. 38 (5): 715–721. doi:10.1007/BF02755923
- Holleman, Arnold F.; Wiberg, Egon; Wiberg, Nils (1985). "Tin". Lehrbuch der Anorganischen Chemie (in German) (91–100 ed.). Walter de Gruyter. pp. 793–800. ISBN 3-11-007511-3
- Korhonen, T; Puska, M.; Nieminen, R. (1995). "Vacancy formation energies for fcc and bcc transition metals". Phys. Rev. B. 51 (15): 9526–9532. Bibcode:1995PhRvB..51.9526K. doi:10.1103/PhysRevB.51.9526

2

Classification of Materials

Materials can be normally classified into metals, ceramics and polymers, and each of them can be further subdivided into smaller sub-groups. Metals can be further simplified into ferrous and non-ferrous materials. Ceramics can be categories into refractory materials, glass, abrasives, and advanced ceramics. Materials are best understood in confluence with the major topics listed in the following section.

Material

Material is a broad term for a chemical substance or mixture of substances that constitute a thing.

In the metaphysical sense, materials can be anything something else is consisting of, whether pure or impure, a singular composite or a complex mix, living or non-living matter, whether natural or man-made, either concrete or abstract. Materials are given classifications based on the different properties defined as physical, chemical, geological, biological, choreographical or philosophical. In the physical sense, materials are studied in materials science.

In the industrial sense, materials are inputs to production or manufacturing processes. They may either be raw material, that is, unprocessed, or processed before being used in more advanced production processes, either by destillation or synthesis (synthetic materials).

Classification of Materials

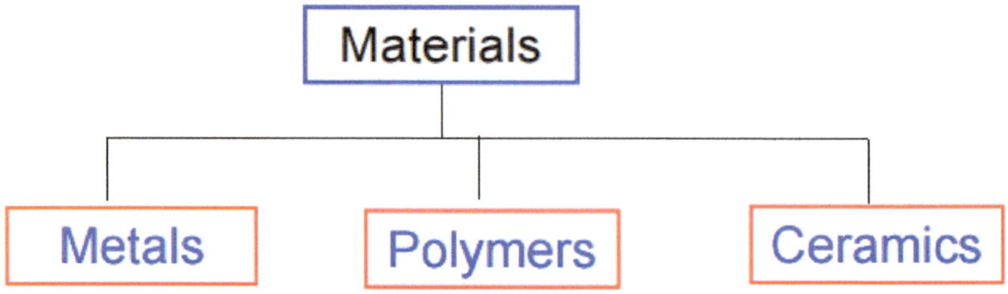

Types of materials include:

- Biomaterial, of biological origin

- Composite material, composed of multiple materials with differing physical properties
- Textiles, sometimes referred to as "material"
- Genetic material.

Metal

A metal is a material (an element, compound, or alloy) that is typically hard, opaque, shiny, and has good electrical and thermal conductivity. Metals are generally malleable—that is, they can be hammered or pressed permanently out of shape without breaking or cracking—as well as fusible (able to be fused or melted) and ductile (able to be drawn out into a thin wire). About 91 of the 118 elements in the periodic table are metals, the others are nonmetals or metalloids. Some elements appear in both metallic and non-metallic forms.

Astrophysicists use the term "metal" to collectively describe all elements other than hydrogen and helium, the simplest two, in a star. The star fuses smaller atoms, mostly hydrogen and helium, to make larger ones over its lifetime. In that sense, the metallicity of an object is the proportion of its matter made up of all heavier chemical elements, not just traditional metals.

Many elements and compounds that are not normally classified as metals become metallic under high pressures; these are formed as metallic allotropes of non-metals.

Structure and Bonding

The atoms of metallic substances are typically arranged in one of three common crystal structures, namely body-centered cubic (bcc), face-centered cubic (fcc), and hexagonal close-packed (hcp). In bcc, each atom is positioned at the center of a cube of eight others. In fcc and hcp, each atom is surrounded by twelve others, but the stacking of the layers differs. Some metals adopt different structures depending on the temperature.

Atoms of metals readily lose their outer shell electrons, resulting in a free flowing cloud of electrons within their otherwise solid arrangement. This provides the ability of metallic substances to easily transmit heat and electricity. While this flow of electrons occurs, the solid characteristic of the metal is produced by electrostatic interactions between each atom and the electron cloud. This type of bond is called a metallic bond.

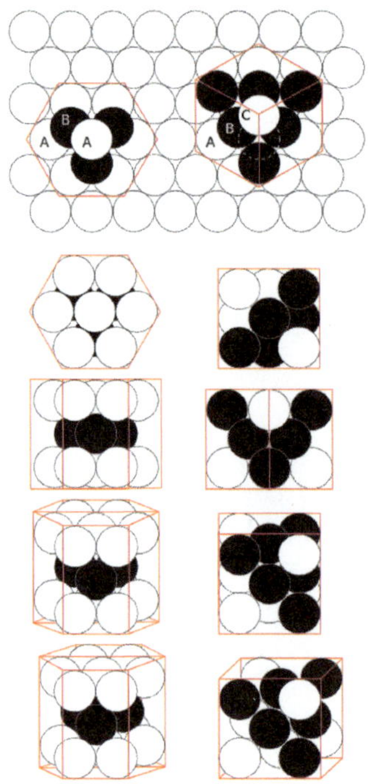

hcp and fcc close-packing of spheres

Properties

Chemical

Metals are usually inclined to form cations through electron loss, reacting with oxygen in the air to form oxides over various timescales (iron rusts over years, while potassium burns in seconds). Examples:

$$4\,Na + O_2 \rightarrow 2\,Na_2O \text{ (sodium oxide)}$$

$$2\,Ca + O_2 \rightarrow 2\,CaO \text{ (calcium oxide)}$$

$$4\,Al + 3\,O_2 \rightarrow 2\,Al_2O_3 \text{ (aluminium oxide).}$$

The transition metals (such as iron, copper, zinc, and nickel) are slower to oxidize because they form a passivating layer of oxide that protects the interior. Others, like palladium, platinum and gold, do not react with the atmosphere at all. Some metals form a barrier layer of oxide on their surface which cannot be penetrated by further oxygen molecules and thus retain their shiny appearance and good conductivity for many decades (like aluminium, magnesium, some steels, and titanium). The oxides of metals are generally basic, as opposed to those of nonmetals, which are acidic. Exceptions are largely oxides with very high oxidation states such as CrO_3, Mn_2O_7, and OsO_4, which have strictly acidic reactions.

Painting, anodizing or plating metals are good ways to prevent their corrosion. However, a more reactive metal in the electrochemical series must be chosen for coating, especially when chipping of the coating is expected. Water and the two metals form an electrochemical cell, and if the coating is less reactive than the coatee, the coating actually *promotes* corrosion.

Physical

Metals in general have high electrical conductivity, high thermal conductivity, and high density. Typically they are malleable and ductile, deforming under stress without cleaving. In terms of optical properties, metals are shiny and lustrous. Sheets of metal beyond a few micrometres in thickness appear opaque, but gold leaf transmits green light.

Gallium crystals

Although most metals have higher densities than most nonmetals, there is wide variation in their densities, lithium being the least dense solid element and osmium the densest. The alkali and alkaline earth metals in groups I A and II A are referred to as the light metals because they have low density, low hardness, and low melting points. The high density of most metals is due to the tightly packed crystal lattice of the metallic structure. The strength of metallic bonds for different metals reaches a maximum around the center of the transition metal series, as those elements have large amounts of delocalized electrons in tight binding type metallic bonds. However, other factors (such as atomic radius, nuclear charge, number of bonds orbitals, overlap of orbital energies and crystal form) are involved as well.

Electrical

The electrical and thermal conductivities of metals originate from the fact that their outer electrons are delocalized. This situation can be visualized by seeing the atomic structure of a metal as a collection of atoms embedded in a sea of highly mobile electrons. The electrical conductivity, as well as the electrons' contribution to the heat capacity and heat conductivity of metals can be calculated from the free electron model, which does not take into account the detailed structure of the ion lattice.

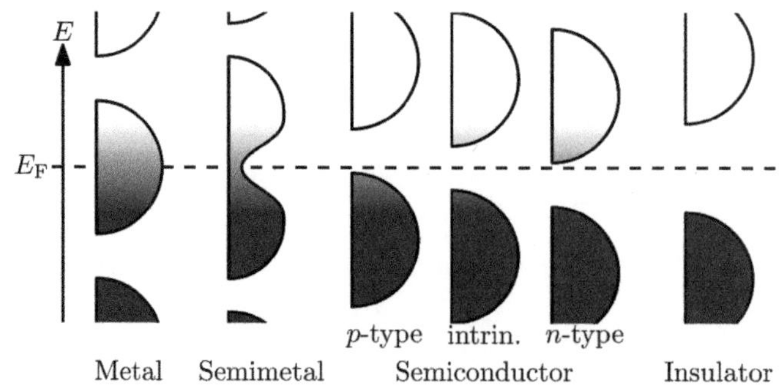

Filling of the electronic states in various types of materials at equilibrium. Here, height is energy while width is the density of available states for a certain energy in the material listed. The shade follows the Fermi–Dirac distribution (black = all states filled, white = no state filled). In metals and semimetals the Fermi level E_F lies inside at least one band. In insulators and semiconductors the Fermi level is inside a band gap; however, in semiconductors the bands are near enough to the Fermi level to be thermally populated with electrons or holes.

When considering the electronic band structure and binding energy of a metal, it is necessary to take into account the positive potential caused by the specific arrangement of the ion cores—which is periodic in crystals. The most important consequence of the periodic potential is the formation of a small band gap at the boundary of the Brillouin zone. Mathematically, the potential of the ion cores can be treated by various models, the simplest being the nearly free electron model.

Mechanical

Mechanical properties of metals include ductility, i.e. their capacity for plastic deformation. Reversible elastic deformation in metals can be described by Hooke's Law for restoring forces, where the stress is linearly proportional to the strain. Forces larger than the elastic limit, or heat, may cause a permanent (irreversible) deformation of the object, known as plastic deformation or plasticity. This irreversible change in atomic arrangement may occur as a result of:

- The action of an applied force (or work). An applied force may be tensile (pulling) force, compressive (pushing) force, shear, bending or torsion (twisting) forces.
- A change in temperature (heat). A temperature change may affect the mobility of the structural defects such as grain boundaries, point vacancies, line and screw dislocations, stacking faults and twins in both crystalline and non-crystalline solids. The movement or displacement of such mobile defects is thermally activated, and thus limited by the rate of atomic diffusion.

Viscous flow near grain boundaries, for example, can give rise to internal slip, creep and fatigue in metals. It can also contribute to significant changes in the microstructure like grain growth and localized densification due to the elimination of intergranular porosity. Screw dislocations may slip in the direction of any lattice plane containing the dislocation, while the principal driving force for "dislocation climb" is the movement or diffusion of vacancies through a crystal lattice.

Hot metal work from a blacksmith.

In addition, the nondirectional nature of metallic bonding is also thought to contribute significantly to the ductility of most metallic solids. When the planes of an ionic bond slide past one another, the resultant change in location shifts ions of the same charge into close proximity, resulting in the cleavage of the crystal; such shift is not observed in covalently bonded crystals where fracture and crystal fragmentation occurs.

Alloys

An alloy is a mixture of two or more elements in which the main component is a metal. Most pure metals are either too soft, brittle or chemically reactive for practical use. Combining different ratios of metals as alloys modifies the properties of pure metals to produce desirable characteristics. The aim of making alloys is generally to make them less brittle, harder, resistant to corrosion, or have a more desirable color and luster. Of all the metallic alloys in use today, the alloys of iron (steel, stainless steel, cast iron, tool steel, alloy steel) make up the largest proportion both by quantity and commercial value. Iron alloyed with various proportions of carbon gives low, mid and high carbon steels, with increasing carbon levels reducing ductility and toughness. The addition of silicon will produce cast irons, while the addition of chromium, nickel and molybdenum to carbon steels (more than 10%) results in stainless steels.

Other significant metallic alloys are those of aluminium, titanium, copper and magnesium. Copper alloys have been known since prehistory—bronze gave the Bronze Age its name—and have many applications today, most importantly in electrical wiring. The alloys of the other three metals have been developed relatively recently; due to their chemical reactivity they require electrolytic extraction processes. The alloys of aluminium, titanium and magnesium are valued for their high strength-to-weight ratios; magnesium can also provide electromagnetic shielding. These materials are ideal for situations where high strength-to-weight ratio is more important than material cost, such as in aerospace and some automotive applications.

Alloys specially designed for highly demanding applications, such as jet engines, may contain more than ten elements.

Categories

Base Metal

Zinc, a base metal, reacting with an acid

In chemistry, the term *base metal* is used informally to refer to a metal that is easily oxidized or corroded, and reacts easily with dilute hydrochloric acid (HCl) to form hydrogen. Examples include iron, nickel, lead and zinc. Copper is considered a base metal as it is oxidized relatively easily, although it does not react with HCl. Base metal is commonly used in opposition to noble metal.

In alchemy, a *base metal* was a common and inexpensive metal, as opposed to precious metals, mainly gold and silver. A longtime goal of the alchemists was the transmutation of base metals into precious metals.

In numismatics, coins in the past derived their value primarily from the precious metal content. Most modern currencies are fiat currency, allowing the coins to be made of base metal.

Ferrous Metal

The term "ferrous" is derived from the Latin word meaning "containing iron". This can include pure iron, such as wrought iron, or an alloy such as steel. Ferrous metals are often magnetic, but not exclusively.

Noble Metal

Noble metals are metals that are resistant to corrosion or oxidation, unlike most base metals. They tend to be precious metals, often due to perceived rarity. Examples include gold, platinum, silver, rhodium, iridium and palladium.

Precious Metal

A gold nugget

A *precious metal* is a rare metallic chemical element of high economic value.

Chemically, the precious metals are less reactive than most elements, have high luster and high electrical conductivity. Historically, precious metals were important as currency, but are now regarded mainly as investment and industrial commodities. Gold, silver, platinum and palladium each have an ISO 4217 currency code. The best-known precious metals are gold and silver. While both have industrial uses, they are better known for their uses in art, jewelry, and coinage. Other precious metals include the platinum group metals: ruthenium, rhodium, palladium, osmium, iridium, and platinum, of which platinum is the most widely traded.

The demand for precious metals is driven not only by their practical use, but also by their role as investments and a store of value. Palladium was, as of summer 2006, valued at a little under half the price of gold, and platinum at around twice that of gold. Silver is substantially less expensive than these metals, but is often traditionally considered a precious metal for its role in coinage and jewelry.

Heavy Metal

A heavy metal is any relatively dense metal or metalloid. More specific definitions have been proposed, but none have obtained widespread acceptance. Some heavy metals have niche uses, or are notably toxic; some are essential in trace amounts.

Extraction

Metals are often extracted from the Earth by means of mining ores that are rich sources of the requisite elements, such as bauxite. Ore is located by prospecting techniques, followed by the exploration and examination of deposits. Mineral sources are generally divided into surface mines, which are mined by excavation using heavy equipment, and subsurface mines.

Once the ore is mined, the metals must be extracted, usually by chemical or electrolytic reduction. Pyrometallurgy uses high temperatures to convert ore into raw metals, while hydrometallurgy employs aqueous chemistry for the same purpose. The methods used depend on the metal and their contaminants.

When a metal ore is an ionic compound of that metal and a non-metal, the ore must usually be smelted—heated with a reducing agent—to extract the pure metal. Many common metals, such as iron, are smelted using carbon as a reducing agent. Some metals, such as aluminium and sodium, have no commercially practical reducing agent, and are extracted using electrolysis instead.

Sulfide ores are not reduced directly to the metal but are roasted in air to convert them to oxides.

Recycling

Demand for metals is closely linked to economic growth. During the 20th century, the variety of metals uses in society grew rapidly. Today, the development of major nations, such as China and India, and advances in technologies, are fuelling ever more demand. The result is that mining activities are expanding, and more and more of the world's metal stocks are above ground in use, rather than below ground as unused reserves. An example is the in-use stock of copper. Between 1932 and 1999, copper in use in the US rose from 73g to 238g per person.

Metals are inherently recyclable, so in principle, can be used over and over again, minimizing these negative environmental impacts and saving energy. For example, 95% of the energy used to make aluminium from bauxite ore is saved by using recycled material. Levels of metals recycling are generally low. In 2010, the International Resource Panel, hosted by the United Nations Environment Programme (UNEP) published reports on metal stocks that exist within society and their recycling rates.

The report authors observed that the metal stocks in society can serve as huge mines above ground. They warned that the recycling rates of some rare metals used in applications such as mobile phones, battery packs for hybrid cars and fuel cells are so low that unless future end-of-life recycling rates are dramatically stepped up these critical metals will become unavailable for use in modern technology.

Metallurgy

Metallurgy is a domain of materials science that studies the physical and chemical behavior of metallic elements, their intermetallic compounds, and their mixtures, which are called alloys.

Applications

Some metals and metal alloys possess high structural strength per unit mass, making them useful materials for carrying large loads or resisting impact damage. Metal alloys can be engineered to have high resistance to shear, torque and deformation. However the same metal can also be vulnerable to fatigue damage through repeated use or from sudden stress failure when a load capacity is exceeded. The strength and resilience of metals has led to their frequent use in high-rise building and bridge construction, as well as most vehicles, many appliances, tools, pipes, non-illuminated signs and railroad tracks.

The two most commonly used structural metals, iron and aluminium, are also the most abundant metals in the Earth's crust.

Metals are good conductors, making them valuable in electrical appliances and for carrying an electric current over a distance with little energy lost. Electrical power grids rely on metal cables to distribute electricity. Home electrical systems, for the most part, are wired with copper wire for its good conducting properties.

The thermal conductivity of metal is useful for containers to heat materials over a flame. Metal is also used for heat sinks to protect sensitive equipment from overheating.

The high reflectivity of some metals is important in the construction of mirrors, including precision astronomical instruments. This last property can also make metallic jewelry aesthetically appealing.

Some metals have specialized uses; radioactive metals such as uranium and plutonium are used in nuclear power plants to produce energy via nuclear fission. Mercury is a liquid at room temperature and is used in switches to complete a circuit when it flows over the switch contacts. Shape memory alloy is used for applications such as pipes, fasteners and vascular stents.

Metals can be doped with foreign molecules—organic, inorganic, biological and polymers. This doping entails the metal with new properties that are induced by the guest molecules. Applications in catalysis, medicine, electrochemical cells, corrosion and more have been developed.

Trade

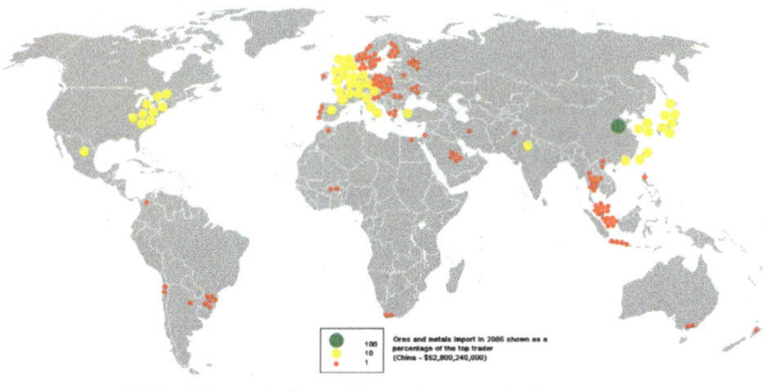

World metal and ore imports in 2005

The World Bank reports that China was the top importer of ores and metals in 2005 followed by the United States and Japan.

History

The nature of metals has fascinated humans for many centuries, because these materials provided people with tools of unsurpassed properties both in war and in their preparation and processing. Pure gold and silver were known to humans since the Stone Age. Lead and silver were fused from their ores as early as the fourth millennium BC.

Ancient Latin and Greek writers such as Theophrastus, Pliny the Elder in his *Natural History*, or Pedanius Dioscorides, did not try to classify metals. The ancient Europeans never attained the concept "metal" as a distinct elementary substance with fixed, characteristic chemical and physical properties. Following Empedocles, all substances within the sublunary sphere were assumed to vary in their constituent classical elements of earth, water, air and fire. Following the Pythagoreans, Plato assumed that these elements could be further reduced to plane geometrical shapes (triangles and squares) bounding space and relating to the regular polyhedra in the sequence earth:cube, water:icosahedron, air:octahedron, fire:tetrahedron. However, this philosophical extension did not become as popular as the simple four elements, after it was rejected by Aristotle. Aristotle also rejected the atomic theory of Democritus, since he classified the implied existence of a vacuum necessary for motion as a contradiction (a vacuum implies nonexistence, therefore cannot exist). Aristotle did, however, introduce underlying antagonistic qualities (or forces) of dry vs. wet and cold vs. heat into the composition of each of the four elements. The word "metal" originally meant "mines" and only later gained the general meaning of products from materials obtained in mines. In the first centuries A.D. a relation between the planets and the existing metals was assumed as Gold:Sun, Silver:Moon, Electrum:Jupiter, Iron:Mars, Copper:Venus, Tin:Mercury, Lead:Saturn. After electrum was determined to be a combination of silver and gold, the relations Tin:Jupiter and Mercury:Mercury were substituted into the previous sequence.

Arabic and medieval alchemists believed that all metals, and in fact, all sublunar matter, were composed of the principle of sulfur, carrying the combustible property, and the principle of mercury, the mother of all metals and carrier of the liquidity or fusibility, and the volatility properties. These principles were not necessarily the common substances sulfur and mercury found in most laboratories. This theory reinforced the belief that the all metals were destined to become gold in the bowels of the earth through the proper combinations of heat, digestion, time, and elimination of contaminants, all of which could be developed and hastened through the knowledge and methods of alchemy. Paracelsus added the third principle of salt, carrying the nonvolatile and incombustible properties, in his *tria prima* doctrine. These theories retained the four classical elements as underlying the composition of sulfur, mercury and salt.

The first systematic text on the arts of mining and metallurgy was *De la Pirotechnia* by Vannoccio Biringuccio, which treats the examination, fusion, and working of metals. Sixteen years later, Georgius Agricola published *De Re Metallica* in 1555, a clear and complete account of the profession of mining, metallurgy, and the accessory arts and sciences, as well as qualifying as the greatest treatise on the chemical industry through the sixteenth century. He gave the following description of a metal in his *De Natura Fossilium* (1546).

Metal is a mineral body, by nature either liquid or somewhat hard. The latter may be melted by the heat of the fire, but when it has cooled down again and lost all heat, it becomes hard again and resumes its proper form. In this respect it differs from the stone which melts in the fire, for although the latter regain its hardness, yet it loses its pristine form and properties. Traditionally there are six different kinds of metals, namely gold, silver, copper, iron, tin and lead. There are really others, for quicksilver is a metal, although the Alchemists disagree with us on this subject, and bismuth is also. The ancient Greek writers seem to have been ignorant of bismuth, wherefore Ammonius rightly states that there are many species of metals, animals, and plants which are unknown to us. Stibium when smelted in the crucible and refined has as much right to be regarded as a proper metal as is accorded to lead by writers. If when smelted, a certain portion be added to tin, a bookseller's alloy is produced from which the type is made that is used by those who print books on paper. Each metal has its own form which it preserves when separated from those metals which were mixed with it. Therefore neither electrum nor Stannum [not meaning our tin] is of itself a real metal, but rather an alloy of two metals. Electrum is an alloy of gold and silver, Stannum of lead and silver. And yet if silver be parted from the electrum, then gold remains and not electrum; if silver be taken away from Stannum, then lead remains and not Stannum. Whether brass, however, is found as a native metal or not, cannot be ascertained with any surety. We only know of the artificial brass, which consists of copper tinted with the colour of the mineral calamine. And yet if any should be dug up, it would be a proper metal. Black and white copper seem to be different from the red kind. Metal, therefore, is by nature either solid, as I have stated, or fluid, as in the unique case of quicksilver. But enough now concerning the simple kinds.

Stainless Steel

- Stainless steels - A group of steels that contain at least 11% Cr. Exhibits extraordinary corrosion resistance due to formation of a very thin layer of Cr_2O_3 on the surface.

Categories of stainless steels:

- Ferritic Stainless Steels – Composed of α ferrite (BCC)
- Martensitic Stainless Steels – Can be heat treated.
- Austenitic Stainless Steels – Austenite (γ) phase field is extended to room temperature. Most corrosion resistant.

- Precipitation-Hardening (PH) Stainless Steels – Ultra high- strength due to precipitation hardening.
- Duplex Stainless Steels – Ferrite + Austenite

Applications of Stainless Steels

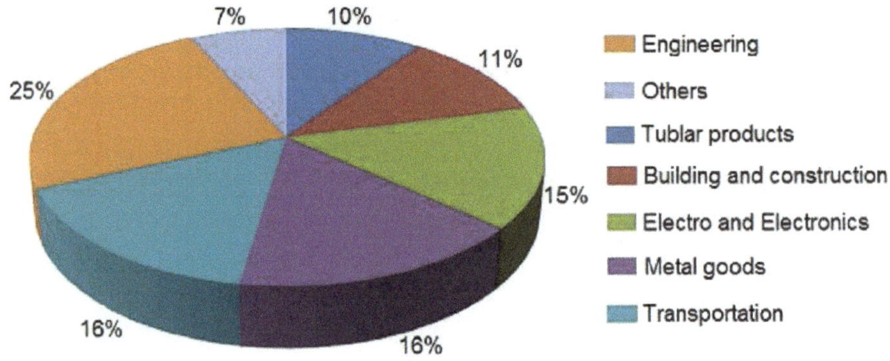

Nonferrous Metals

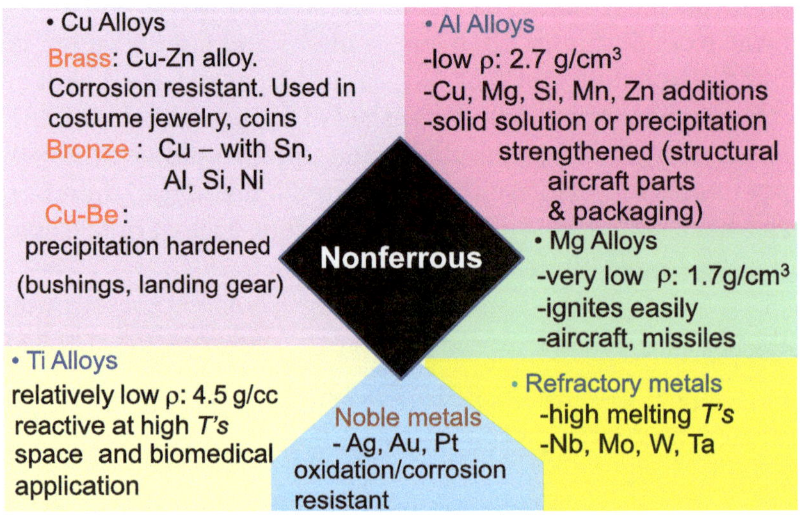

Copper

- Copper is one of the earliest metals discovered by man.
- The boilers on early steamboats were made from copper.
- The copper tubing used in water plumbing in Pyramids was found in serviceable condition after more than 5,000 years.

- Cu is a ductile metal. Pure Cu is soft and malleable, difficult to machine.
- Very high electrical conductivity – second only to silver.
- Copper is refined to high purity for many electrical applications.
- Excellent thermal conductivity – Copper cookware most highly regarded – fast and uniform heating.
- Electrical and construction industries are the largest users of Cu.
- The second largest use of Cu is probably in coins.
- The U.S. nickel is actually 75% copper. The dime, quarter, and half dollar coins contain 91.67% copper and the Susan B Anthony dollar is 87.5% copper.
- The various Euro coins are made of Cu-Ni, Cu-Zn-Ni or Cu-Al-Zn-Sn alloys.

Copper Alloys

- Brasses and Bronzes are most commonly used alloys of Cu. Brass is an alloy with Zn. Bronzes contain tin, aluminum, silicon or beryllium.
- Other copper alloy families include copper-nickels and nickel silvers. More than 400 copper-base alloys are recognized.

Family of Cu Alloys		
Alloy	**Alloying element**	**UNS numbers**
Brass	Zinc (Zn)	C1xxxx–C4xxxx,C66400– C69800
Phosphor bronze	Tin (Sn)	C5xxxx
Aluminium bronzes	Aluminium (Al)	C60600–C64200
Silicon bronzes	Silicon (Si)	C64700–C66100
Copper nickel, nickel silvers	Nickel (Ni)	C7xxxx

Copper Alloys – Brass

- Brass is the most common alloy of Cu – It's an alloy with Zn
- Brass has higher ductility than copper or zinc.
- Easy to cast - Relatively low melting point and high fluidity
- Properties can be tailored by varying Zn content.
- Some of the common brasses are yellow, naval and cartridge
- Brass is frequently used to make musical instruments (good ductility and acoustic properties).

Bronze

- Copper alloys containing tin, lead, aluminum, silicon and nickel are classified as bronzes.
- Cu-Sn Bronze is one of the earliest alloy to be discovered as Cu ores invariably contain Sn.
- Stronger than brasses with good corrosion and tensile properties; can be cast, hot worked and cold worked.
- Wide range of applications: ancient Chinese cast artifacts, skateboard ball bearings, surgical and dental instruments.

Beryllium Copper

- Cu-Be alloys are heat treatable. Max solubility of Be in Cu is 2.7% at 866 °C. Decreasing solubility at lower temp. imparts precipitation hardening ability.
- Cast alloys - higher Be. Wrought alloys – lower Be and some Co

- Cu-Be is ductile, weldable and machinable. Also resistant to non-oxidizing acids (HCl or H_2CO_3), abrasive wear and galling.
- Thermal conductivity is between steels and aluminum.

Applications

- Used in springs, load cells and other parts subjected to repeated loading. Low-current contacts for batteries and electrical connectors. Cast alloys are used in injection molds. Other applications include jet aircraft landing gear bearings and bushings and percussion instruments.

Compositions, Properties and Application of Some Cu Alloys

Name	UNS No.	Compos. (wt.%)	Condition	YS (MPa)	UTS (MPa)	%El	Applications
Eletrolytic Copper	C11000	0.04 O	Annealed	69	220	45	Electrical wires, roofing, nails, rivets
Cartridge brass	C26000	30.0 Zn	Cold- rolled	435	525	8	Automotive radiator core, lamp fixture, ammunition.
Phosphor bronze	C51000	5.0 Sn, 0.25 P	Annealed	130	325	64	Bellows, clutch disk, diaphragm, fuse clips, springs
Yellow bras (Leaded)	C85400	29 Zn, 3.0 Pb, 1 Sn	As cast	83	234	35	Furniture, radiator fittings, battery clamps, light fixtures
Al bronze	C95400	11 Al, 4 Fe	As cast	241	586	18	Bearings, bushings, valve seats and guards
Beryllium copper	C17200	1.9 Be 0.2 Co	Precipita. hardened	965	1140	10	Electrical, valves, pumps
Cu - Ni	C71500	30 Ni	Annealed	125	380	36	Condenser, heat-exchanger, piping, valves
Tin bronze	C90500	10 Sn, 2 Zn	As cast	152	310	25	Bearings, bushing, piston rings, gears

Aluminum

- Aluminum is a light metal (ρ = 2.7 g/cc); is easily machinable; has wide variety of surface finishes; good electrical and thermal conductivities; highly reflective to heat and light.
- Versatile metal - can be cast, rolled, stamped, drawn, spun, roll-formed, hammered, extruded and forged into many shapes.

- Aluminum can be riveted, welded, brazed, or resin bonded.
- Corrosion resistant - no protective coating needed, however it is often anodized to improve surface finish, appearance.
- Al and its alloys - high strength-to-weight ratio (high specific strength) owing to low density.
- Such materials are widely used in aerospace and automotive applications where weight savings are needed for better fuel efficiency and performance.
- Al-Li alloys are lightest among all Al alloys and find wide applications in the aerospace industry.

Aluminum Alloys

- Aluminum alloys are classified into two categories – Cast and Wrought alloys.
- Wrought alloys can be either heat-treatable or non-heat treatable.
- Alloys are designated by a 4 digit number. Wrought – the 1st digit indicates the major alloying element. Cast – The last digit after the decimal indicates product from(casting - 0 or ingot -1)

Wrought

Alloy Series	*Principal Alloying Element*
1xxx	Minimum 99.00% Aluminum
2xxx	Copper
3xxx	Manganese
4xxx	Silicon
5xxx	Magnesium
6xxx	Magnesium and Silicon
7xxx	Zinc
8xxx	Other Elements

As Cast

Alloy Series	*Principal Alloying Element*
1xx.x	Aluminum, 99.00% or greater
2xx.x	Copper
3xx.x	Silicon with Copper and/or Magnesium

4xx.x	Silicon
5xx.x	Magnesium
6xx.x	Unused Series
7xx.x	Zinc
8xx.x	Tin
9xx.x	Other Elements

Temper Designations

- F As fabricated - products in which no thermal treatments or strain-hardening.
- H Strain-hardened (wrought products) – strain hardened with or without additional thermal treatment.
- H1 Strain-hardened only - strain-hardened without thermal treatment.
- O Annealed, recrystallized
- T Thermally treated with or without strain-hardening to produce stable tempers other than F, O or H.
- T3 Solution heat-treated and then cold worked.

Solution heat treatment – heating to the single phase region and isothermal holding.

Compositions, Properties and Application of Some Al Alloys

Al Ass. No	Composition (wt.%)	Condition	YS (MPa)	UTS (MPa)	%El	Applications
1100	0.12 Cu	Annealed (O)	35	90	45	Food/chemical handling equipment, heat exchangers light reflectors
3003	0.12 Cu,1.2 Mn, 0.1 Zn	Annealed	40	110	30	Utensils, pressure vessels and piping
5052	2.5 Mg, 0.25 Cr	Strain-hardn. (H32)	195	230	14	Bellows, clutch disk, diaphragm, fuse clips, springs
2024	4.4 Cu, 1.5 Mg, 0.6 Mn	Heat treated (T4)	325	470	20	Aircraft structure, rivets, truck wheels, screw
6061	1 Mg, 0.6 Si, 0.3 Cu, 0.2 Cr	T4	145	240	22	Trucks, canoes, railroad cars, furniture, pipelines
7075	5.6 Zn, 2.5 Mg, 1.6 Cu, 0.23	Peak-aged (T6)	505	570	11	Aircraft structures and other highly loaded applications
359.0	Cr 7 Si, 0.3 Mg	T6	164	228	4	Aircraft pump parts, automotive transmission cases, cylinder blocks

8090	2.0 Li. 1.3 Cu, 0.95 Mg, 0.12 Zr	Heat treated cold-worked (T651)	360	465		Damage tolerant aircraft structures

Titanium

- Pure titanium melts at 1670°C and has a low density of 4.51 g/cc (40% lighter than steel and 60% heavier than aluminum).
- Titanium has high affinity to oxygen – strong deoxidiser. Can catch fire and cause severe damage
- Ti is stronger than Al - high strength and low weight makes titanium very useful as a structural metal.
- Excellent corrosion resistance due to a presence of a protective thin oxide surface film. Can be used as biomaterial.
- Can be used in elevated temperature components.
- Limitation of pure Ti is its lower strength. Alloying is done to improve strength.
- Oxygen, nitrogen, and hydrogen can cause titanium to become more brittle. Care should be taken during processing.
- Titanium can also be cast using a vacuum furnace.
- Because of its high strength to weight ratio and excellent corrosion resistance, titanium is used in a variety of applications:
- Aircraft – Body structure, Engine parts
- sporting equipment, chemical processing, desalination, turbine engine parts, valve and pump parts, marine hardware
- Medical implants - prosthetic devices.
- Recently use of Ti in bikes and automotives is increasing

Titanium Alloys

- Pure Ti exhibits two phases – Hexagonal α-phase at room temperature and BCC β-phase above 882 °C.
- Strength of Titanium is improved by alloying. Alloying elements are either α or β stabilizer.
- Elements with electron/atoms ratio < 4 – α stabilizer (Al, O, Ga), = 4 – neutral (Sn, Zr) and > 4 – β stabilizer (V, Mo,Ta, W).

- (α + β) two-phase alloys can be obtained with right proportions of alloying elements.
- α alloys have low density, moderate strength, reasonable ductility and good creep resistance.
- Metastable β alloys are heavier, stronger and less ductile than α alloys. Creep strength reduces with increasing β content
- (α + β) alloys show a good strength-ductility combination

Compositions, Properties and Application of Some Ti Alloys

Alloy type	Comp . wt.% (UNS No)	Condition	YS (MPa)	UTS (MPa)	%El	Applications
CP Ti	99.1Ti (R50500)	Annealed	414	484	45	Airframe skins, marine and chemical processing equipments
α	Ti-5Al-2.5Sn (R54520)	Annealed	784	826	16	Gas turbine engine casing and rings, chemical processing equipment
Near-α	Ti-8Al-1Mo- 1V (R54810)	Annealed	890	950	15	Forged jet engine components – compressor disc, plate, hubs
α-β	Ti-6Al-4V (R56400)	Annealed	877	947	14	Prosthetic implants, airframe components
α-β	Ti-6Al-6V - 2Sn (R56620)	Annealed	985	1050	14	Rocket engine case, airframe structure
β	Ti-10V -2Fe- 3Al	Heat treated (aging)	1150	1223	10	High-strength airframe components, parts requiring uniform tensile

Nickel

- Nickel is a high-density, high-strength metal with good ductility and excellent corrosion resistance and high temperature properties.
- Ni has may unique properties including its excellent catalytic property. Nickel Catalyst for Fuel Cells: Nickel- cobalt is seen as a low-cost substitute for platinum catalysts.
- Two-thirds of all nickel produced goes into stainless steel production. Also used extensively in electroplating various parts in variety of applications.
- Ni-base super alloys are a unique class of materials having exceptionally good

high temperature strength, creep and oxidation resistance. Used in many high temperature applications like turbine engines.

- Shape Memory Alloys: Ni base (Ni-Ti) and Ni containing (Cu-Al-Ni) shape memory alloys that can go back to original form, are an important class of engineering materials finding widespread use in many applications.
- Nickel-containing materials are used in buildings and infrastructure, chemical production, communications, energy (batteries: Ni-Cd, Ni-metal hydrides), environmental protection, food preparation, water treatment and transportation.

Applications of Nickel

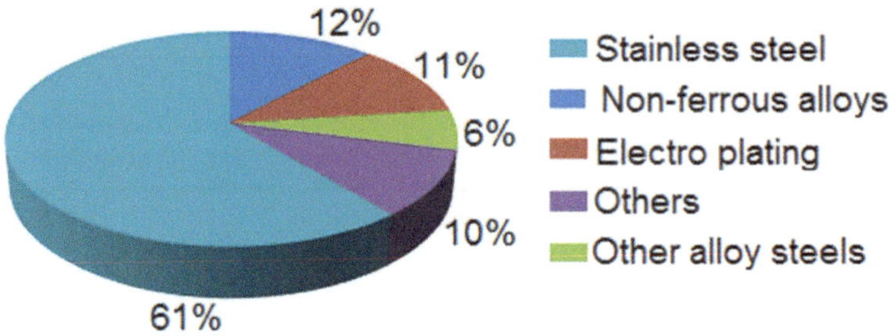

Turbine engine

Electroplating

Batteries

Magnesium

- Magnesium - Lightest among commonly used metals (ρ 1.7 g/cm³). Melting point is 650 °C and it has HCP structure.
- Is very reactive and readily combustible in air. Can be used as igniter or fire-starter.
- Thermal conductivity is less than Al while their CTE is almost same.
- Pure Mg has adequate atmospheric resistance and moderate strength.
- Properties of Mg can be improved substantially by alloying.
- Favorable atomic size - Can be alloyed with many elements. Most widely used alloying elements are Al, Zn, Mn and Zr.

- Mg Alloys – Cast, Wrought
- Wrought alloys are available in rod, bar, sheet, plate, forgings and extrusions.

Magnesium Alloys

- Mg alloys: Impact and dent resistant, have good damping capacity - effective for high-speed applications.
- Due to its light weight, superior machinability and ease of casting, Mg and its alloys are used in many applications:– Auto parts, sporting goods, power tools, aerospace equipment, fixtures, electronic gadgets, and material handling equipment.
- Automotive applications include gearboxes, valve covers, alloy wheels, clutch housings, and brake pedal brackets.

Compositions, Properties and Application of some Mg Alloys

ASTM No	Compos. (wt.%)	Condition	YS (MPa)	UTS (MPa)	%El	Applications
			Wrought Alloys			
AZ31B	3.0 Al, 1.0Zn, 0.2 Mn	Extruded	200	262	15	Structure and tubing, cathodic protection
HK31A	3.0Th, 0.6Zr	Strain hard. Annealed	200	255	9	High temp applications (high strength to 315 °C)
ZK60A	5.5 Zn, 0.45Zr	Aged	285	350	11	Forging of max strength for aircrafts
			Cast alloys			
AZ91D	9.0Al, 0.15 Mn, 0.7 Zn	As cast	150	230	3	Die-cast parts for automobile, luggage and electronic devices
AM60A	9.0Al, 0.13Mn,	As cast	130	220	6	Automotive wheels
AS41A	4.3Al, 1.0 Si, 0.35Mn	As cast	140	210	6	Die-cast parts requiring good creep strength

Ceramic

A selection of silicon nitride components.

Mid-16th century ceramic tilework on the Dome of the Rock, Jerusalem

Spherical Hanging Ornament, 1575-1585, Ottoman Period. Brooklyn Museum.

A ceramic is an inorganic, non-metallic, solid material comprising metal, non-metal or metalloid atoms primarily held in ionic and covalent bonds. This article gives an overview of ceramic materials from the point of view of materials science.

The crystallinity of ceramic materials ranges from highly oriented to semi-crystalline, vitrified, and often completely amorphous (e.g., glasses). Most often, fired ceramics are either vitrified or semi-vitrified as is the case with earthenware, stoneware, and porcelain. Varying crystallinity and electron consumption in the ionic and covalent bonds cause most ceramic materials to be good thermal and electrical insulators (extensively researched in ceramic engineering). With such a large range of possible options for the composition/structure of a ceramic (e.g. nearly all of the elements, nearly all types of bonding, and all levels of crystallinity), the breadth of the subject is vast, and identifiable attributes (e.g. hardness, toughness, electrical conductivity, etc.) are hard to specify for the group as a whole. General properties such as high melting temperature, high hardness, poor conductivity, high moduli of elasticity, chemical resistance and low ductility are the norm, with known exceptions to each of these rules (e.g. piezoelectric ceramics, glass transition temperature, superconductive ceramics, etc.). Many composites, such as fiberglass and carbon fiber, while containing ceramic materials, are not considered to be part of the ceramic family.

The earliest ceramics made by humans were pottery objects (i.e. *pots* or *vessels*) or figurines made from clay, either by itself or mixed with other materials like silica, hardened, sintered, in fire. Later ceramics were glazed and fired to create smooth, colored surfaces, decreasing porosity through the use of glassy, amorphous ceramic coatings on top of the crystalline ceramic substrates. Ceramics now include domestic, industrial and building products, as well as a wide range of ceramic art. In the 20th century, new ceramic materials were developed for use in advanced ceramic engineering, such as in semiconductors.

The word "*ceramic*" comes from the Greek word (*keramikos*), "of pottery" or "for pottery", from (*keramos*), "potter's clay, tile, pottery". The earliest known mention of the root "ceram-" is the Mycenaean Greek *ke-ra-me-we*, "workers of ceramics", written in Linear B syllabic script. The word "ceramic" may be used as an adjective to describe a material, product or process, or it may be used as a noun, either singular, or, more commonly, as the plural noun "ceramics".

Types of Ceramic Material

A ceramic material is an inorganic, non-metallic, often crystalline oxide, nitride or carbide material. Some elements, such as carbon or silicon, may be considered ceramics. Ceramic materials are brittle, hard, strong in compression, weak in shearing and tension. They withstand chemical erosion that occurs in other materials subjected to acidic or caustic environments. Ceramics generally can withstand very high temperatures, such as temperatures that range from 1,000 °C to 1,600 °C (1,800 °F to 3,000 °F). Glass is often not considered a ceramic because of its amorphous (noncrystalline) character. However, glassmaking involves several steps of the ceramic process and its mechanical properties are similar to ceramic materials.

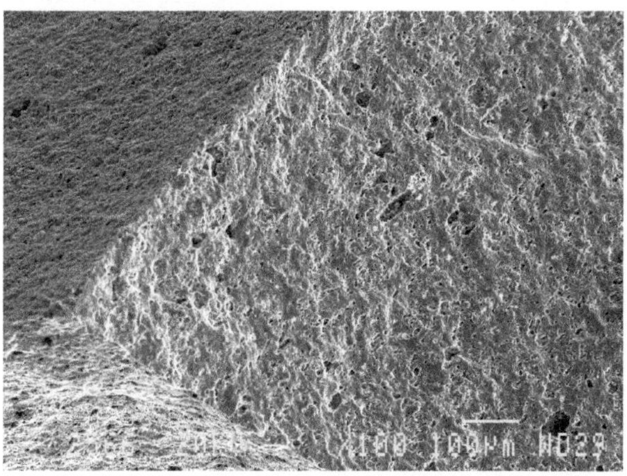

A low magnification SEM micrograph of an advanced ceramic material. The properties of ceramics make fracturing an important inspection method.

Traditional ceramic raw materials include clay minerals such as kaolinite, whereas more recent materials include aluminium oxide, more commonly known as alumina. The modern ceramic materials, which are classified as advanced ceramics, include silicon carbide and tungsten carbide. Both are valued for their abrasion resistance, and hence find use in applications such as the wear plates of crushing equipment in mining operations. Advanced ceramics are also used in the medicine, electrical, electronics industries and body armor.

Crystalline Ceramics

Crystalline ceramic materials are not amenable to a great range of processing. Methods for dealing with them tend to fall into one of two categories – either make the ceramic in the desired shape, by reaction *in situ*, or by "forming" powders into the desired shape, and then sintering to form a solid body. Ceramic forming techniques include shaping by hand (sometimes including a rotation process called "throwing"), slip casting, tape casting (used for making very thin ceramic capacitors, e.g.), injection molding, dry pressing, and other variations. Details of these processes are described in the two books listed below. A few methods use a hybrid between the two approaches.

Noncrystalline Ceramics

Noncrystalline ceramics, being glass, tend to be formed from melts. The glass is shaped when either fully molten, by casting, or when in a state of toffee-like viscosity, by methods such as blowing into a mold. If later heat treatments cause this glass to become partly crystalline, the resulting material is known as a glass-ceramic, widely used as cook-top and also as a glass composite material for nuclear waste disposal.

Properties of Ceramics

The physical properties of any ceramic substance are a direct result of its crystalline structure and chemical composition. Solid state chemistry reveals the fundamental connection between microstructure and properties such as localized density variations, grain size distribution, type of porosity and second-phase content, which can all be correlated with ceramic properties such as mechanical strength σ by the Hall-Petch equation, hardness, toughness, dielectric constant, and the optical properties exhibited by transparent materials.

Physical properties of chemical compounds which provide evidence of chemical composition include odor, colour, volume, density (mass / volume), melting point, boiling point, heat capacity, physical form at room temperature (solid, liquid or gas), hardness, porosity, and index of refraction.

Ceramography is the art and science of preparation, examination and evaluation of ceramic microstructures. Evaluation and characterization of ceramic microstructures is often implemented on similar spatial scales to that used commonly in the emerging field of nanotechnology: from tens of angstroms (A) to tens of micrometers (µm). This is typically somewhere between the minimum wavelength of visible light and the resolution limit of the naked eye.

The microstructure includes most grains, secondary phases, grain boundaries, pores, micro-cracks, structural defects and hardness microindentions. Most bulk mechanical, optical, thermal, electrical and magnetic properties are significantly affected by the observed microstructure. The fabrication method and process conditions are generally indicated by the microstructure. The root cause of many ceramic failures is evident in the cleaved and polished microstructure. Physical properties which constitute the field of materials science and engineering include the following:

Mechanical Properties

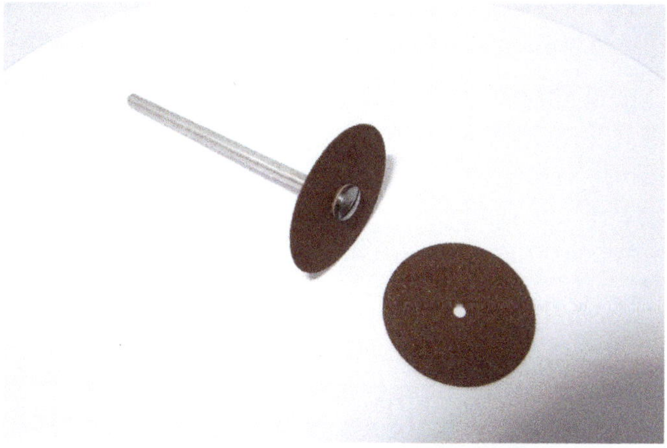

Cutting disks made of silicon carbide

The Porsche Carrera GT's carbon-ceramic (silicon carbide) disc brake

Mechanical properties are important in structural and building materials as well as textile fabrics. They include the many properties used to describe the strength of materials such as: elasticity / plasticity, tensile strength, compressive strength, shear strength, fracture toughness & ductility (low in brittle materials), and indentation hardness.

In modern materials science, fracture mechanics is an important tool in improving the mechanical performance of materials and components. It applies the physics of stress and strain, in particular the theories of elasticity and plasticity, to the microscopic crystallographic defects found in real materials in order to predict the macroscopic mechanical failure of bodies. Fractography is widely used with fracture mechanics to understand the causes of failures and also verify the theoretical failure predictions with real life failures.

Ceramic materials are usually ionic or covalent bonded materials, and can be crystalline or amorphous. A material held together by either type of bond will tend to fracture before any plastic deformation takes place, which results in poor toughness in these materials. Additionally, because these materials tend to be porous, the pores and other microscopic imperfections act as stress concentrators, decreasing the toughness further, and reducing the tensile strength. These combine to give catastrophic failures, as opposed to the normally much more gentle failure modes of metals.

These materials do show plastic deformation. However, due to the rigid structure of the crystalline materials, there are very few available slip systems for dislocations to move, and so they deform very slowly. With the non-crystalline (glassy) materials, viscous flow is the dominant source of plastic deformation, and is also very slow. It is therefore neglected in many applications of ceramic materials.

To overcome the brittle behaviour, ceramic material development has introduced the class of ceramic matrix composite materials, in which ceramic fibers are embedded and with specific coatings are forming fiber bridges across any crack. This mechanism substantially increases the fracture toughness of such ceramics. The ceramic disc brakes are, for example using a ceramic matrix composite material manufactured with a specific process.

Electrical Properties

Semiconductors

Some ceramics are semiconductors. Most of these are transition metal oxides that are II-VI semiconductors, such as zinc oxide.

While there are prospects of mass-producing blue LEDs from zinc oxide, ceramicists are most interested in the electrical properties that show grain boundary effects.

One of the most widely used of these is the varistor. These are devices that exhibit the property that resistance drops sharply at a certain threshold voltage. Once the voltage across the device reaches the threshold, there is a breakdown of the electrical structure in the vicinity of the grain boundaries, which results in its electrical resistance dropping from several megohms down to a few hundred ohms. The major advantage of these is that they can dissipate a lot of energy, and they self-reset – after the voltage across the device drops below the threshold, its resistance returns to being high.

This makes them ideal for surge-protection applications; as there is control over the threshold voltage and energy tolerance, they find use in all sorts of applications. The best demonstration of their ability can be found in electrical substations, where they are employed to protect the infrastructure from lightning strikes. They have rapid response, are low maintenance, and do not appreciably degrade from use, making them virtually ideal devices for this application.

Semiconducting ceramics are also employed as gas sensors. When various gases are passed over a polycrystalline ceramic, its electrical resistance changes. With tuning to the possible gas mixtures, very inexpensive devices can be produced.

Superconductivity

The Meissner effect demonstrated by levitating a magnet above a cuprate superconductor, which is cooled by liquid nitrogen

Under some conditions, such as extremely low temperature, some ceramics exhibit high temperature superconductivity. The exact reason for this is not known, but there are two major families of superconducting ceramics.

Ferroelectricity and Supersets

Piezoelectricity, a link between electrical and mechanical response, is exhibited by a large number of ceramic materials, including the quartz used to measure time in watches and other electronics. Such devices use both properties of piezoelectrics, using electricity to produce a mechanical motion (powering the device) and then using this mechanical motion to produce electricity (generating a signal). The unit of time measured is the natural interval required for electricity to be converted into mechanical energy and back again.

The piezoelectric effect is generally stronger in materials that also exhibit pyroelectricity, and all pyroelectric materials are also piezoelectric. These materials can be used to inter convert between thermal, mechanical, or electrical energy; for instance, after synthesis in a furnace, a pyroelectric crystal allowed to cool under no applied stress generally builds up a static charge of thousands of volts. Such materials are used in motion sensors, where the tiny rise in temperature from a warm body entering the room is enough to produce a measurable voltage in the crystal.

In turn, pyroelectricity is seen most strongly in materials which also display the ferroelectric effect, in which a stable electric dipole can be oriented or reversed by applying an electrostatic field. Pyroelectricity is also a necessary consequence of ferroelectricity. This can be used to store information in ferroelectric capacitors, elements of ferroelectric RAM.

The most common such materials are lead zirconate titanate and barium titanate. Aside from the uses mentioned above, their strong piezoelectric response is exploited in the design of high-frequency loudspeakers, transducers for sonar, and actuators for atomic force and scanning tunneling microscopes.

Positive Thermal Coefficient

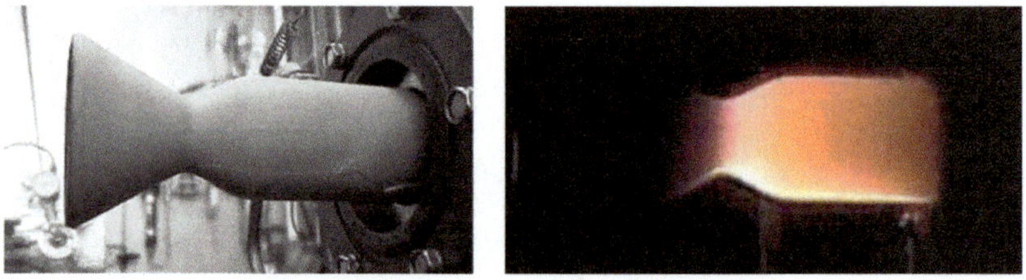

Silicon nitride rocket thruster. Left: Mounted in test stand. Right: Being tested with H_2/O_2 propellants

Increases in temperature can cause grain boundaries to suddenly become insulating in some semiconducting ceramic materials, mostly mixtures of heavy metal titanates. The critical transition temperature can be adjusted over a wide range by variations in chemistry. In such materials, current will pass through the material until joule heating brings it to the transition temperature, at which point the circuit will be broken and current flow will cease. Such ceramics are used as self-controlled heating elements in, for example, the rear-window defrost circuits of automobiles.

At the transition temperature, the material's dielectric response becomes theoretically infinite. While a lack of temperature control would rule out any practical use of the material near its critical temperature, the dielectric effect remains exceptionally strong even at much higher temperatures. Titanates with critical temperatures far below room temperature have become synonymous with "ceramic" in the context of ceramic capacitors for just this reason.

Optical Properties

Cermax xenon arc lamp with synthetic sapphire output window

Optically transparent materials focus on the response of a material to incoming lightwaves of a range of wavelengths. Frequency selective optical filters can be utilized to alter or enhance the brightness and contrast of a digital image. Guided lightwave transmission via frequency selective waveguides involves the emerging field of fiber optics and the ability of certain glassy compositions as a transmission medium for a range of frequencies simultaneously (multi-mode optical fiber) with little or no interference between competing wavelengths or frequencies. This resonant mode of energy and data transmission via electromagnetic (light) wave propagation, though low powered, is virtually lossless. Optical waveguides are used as components in Integrated optical circuits (e.g. light-emitting diodes, LEDs) or as the transmission medium in local and long haul optical communication systems. Also of value to the emerging materials scientist is the sensitivity of materials to radiation in the thermal infrared (IR) portion of the electromagnetic spectrum. This heat-seeking ability is responsible for such diverse optical phenomena as Night-vision and IR luminescence.

Thus, there is an increasing need in the military sector for high-strength, robust materials which have the capability to transmit light (electromagnetic waves) in the visible (0.4 – 0.7 micrometers) and mid-infrared (1 – 5 micrometers) regions of the spectrum. These materials are needed for applications requiring transparent armor, including next-generation high-speed missiles and pods, as well as protection against improvised explosive devices (IED).

In the 1960s, scientists at General Electric (GE) discovered that under the right manufacturing conditions, some ceramics, especially aluminium oxide (alumina), could be made translucent. These translucent materials were transparent enough to be used for containing the electrical plasma generated in high-pressure sodium street lamps. During the past two decades, additional types of transparent ceramics have been developed for applications such as nose cones for heat-seeking missiles, windows for fighter aircraft, and scintillation counters for computed tomography scanners.

In the early 1970s, Thomas Soules pioneered computer modeling of light transmission through translucent ceramic alumina. His model showed that microscopic pores in ceramic, mainly trapped at the junctions of microcrystalline grains, caused light to scatter and prevented true transparency. The volume fraction of these microscopic pores had to be less than 1% for high-quality optical transmission.

This is basically a particle size effect. Opacity results from the incoherent scattering of light at surfaces and interfaces. In addition to pores, most of the interfaces in a typical metal or ceramic object are in the form of grain boundaries which separate tiny regions of crystalline order. When the size of the scattering center (or grain boundary) is reduced below the size of the wavelength of the light being scattered, the scattering no longer occurs to any significant extent.

In the formation of polycrystalline materials (metals and ceramics) the size of the crystalline grains is determined largely by the size of the crystalline particles present in the raw material during formation (or pressing) of the object. Moreover, the size of the grain boundaries scales directly with particle size. Thus a reduction of the original particle size below the wavelength of visible light (~ 0.5 micrometers for shortwave violet) eliminates any light scattering, resulting in a transparent material.

Recently, Japanese scientists have developed techniques to produce ceramic parts that rival the transparency of traditional crystals (grown from a single seed) and exceed the fracture toughness of a single crystal. In particular, scientists at the Japanese firm Konoshima Ltd., a producer of ceramic construction materials and industrial chemicals, have been looking for markets for their transparent ceramics.

Livermore researchers realized that these ceramics might greatly benefit high-powered lasers used in the National Ignition Facility (NIF) Programs Directorate. In particular, a Livermore research team began to acquire advanced transparent ceramics from Konoshima to determine if they could meet the optical requirements needed for Livermore's Solid-State Heat Capacity Laser (SSHCL). Livermore researchers have also been testing applications of these materials for applications such as advanced drivers for laser-driven fusion power plants.

Examples

Until the 1950s, the most important ceramic materials were (1) pottery, bricks and tiles, (2) cements and (3) glass. A composite material of ceramic and metal is known as cermet.

Porcelain high-voltage insulator

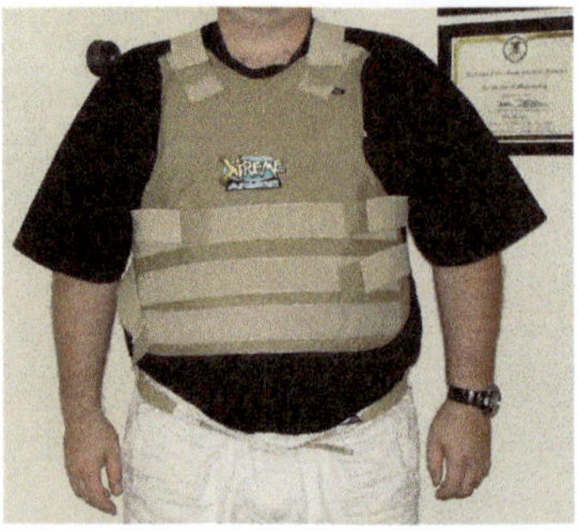

Silicon carbide is used for inner plates of ballistic vests

Other ceramic materials, generally requiring greater purity in their make-up than those above, include forms of several chemical compounds, including:

- Barium titanate (often mixed with strontium titanate) displays ferroelectricity, meaning that its mechanical, electrical, and thermal responses are coupled to one another and also history-dependent. It is widely used in electromechanical transducers, ceramic capacitors, and data storage elements. Grain boundary conditions can create PTC effects in heating elements.
- Bismuth strontium calcium copper oxide, a high-temperature superconductor
- Boron oxide is used in body armor.
- Boron nitride is structurally isoelectronic to carbon and takes on similar physical forms: a graphite-like one used as a lubricant, and a diamond-like one used as an abrasive.
- Earthenware used for domestic ware such as plates and mugs.
- Ferrite is used in the magnetic cores of electrical transformers and magnetic core memory.
- Lead zirconate titanate (PZT) was developed at the United States National Bureau of Standards in 1954. PZT is used as an ultrasonic transducer, as its piezoelectric properties greatly exceed those of Rochelle salt.
- Magnesium diboride (MgB_2) is an unconventional superconductor.
- Porcelain is used for a wide range of household and industrial products.
- Sialon (Silicon Aluminium Oxynitride) has high strength; resistance to thermal shock, chemical and wear resistance, and low density. These ceramics are used in non-ferrous molten metal handling, weld pins and the chemical industry.
- Silicon carbide (SiC) is used as a susceptor in microwave furnaces, a commonly used abrasive, and as a refractory material.
- Silicon nitride (Si_3N_4) is used as an abrasive powder.
- Steatite (magnesium silicates) is used as an electrical insulator.
- Titanium carbide Used in space shuttle re-entry shields and scratchproof watches.
- Uranium oxide (UO_2), used as fuel in nuclear reactors.
- Yttrium barium copper oxide ($YBa_2Cu_3O_{7-x}$), another high temperature superconductor.
- Zinc oxide (ZnO), which is a semiconductor, and used in the construction of varistors.

- Zirconium dioxide (zirconia), which in pure form undergoes many phase changes between room temperature and practical sintering temperatures, can be chemically “stabilized” in several different forms. Its high oxygen ion conductivity recommends it for use in fuel cells and automotive oxygen sensors. In another variant, metastable structures can impart transformation toughening for mechanical applications; most ceramic knife blades are made of this material.
- Partially stabilised zirconia (PSZ) is much less brittle than other ceramics and is used for metal forming tools, valves and liners, abrasive slurries, kitchen knives and bearings subject to severe abrasion.

Kitchen knife with a ceramic blade

Ceramic Products

By Usage

For convenience, ceramic products are usually divided into four main types; these are shown below with some examples:

- Structural, including bricks, pipes, floor and roof tiles
- Refractories, such as kiln linings, gas fire radiants, steel and glass making crucibles
- Whitewares, including tableware, cookware, wall tiles, pottery products and sanitary ware
- Technical, also known as engineering, advanced, special, and fine ceramics. Such items include:
    - gas burner nozzles
    - ballistic protection
    - nuclear fuel uranium oxide pellets
    - biomedical implants
    - coatings of jet engine turbine blades

- o ceramic disk brake
- o missile nose cones
- o bearing (mechanical)
- o tiles used in the Space Shuttle program

Ceramics Made with Clay

Frequently, the raw materials of modern ceramics do not include clays. Those that do are classified as follows:

- Earthenware, fired at lower temperatures than other types
- Stoneware, vitreous or semi-vitreous
- Porcelain, which contains a high content of kaolin
- Bone china

Classification of Technical Ceramics

Technical ceramics can also be classified into three distinct material categories:

- Oxides: alumina, beryllia, ceria, zirconia
- Nonoxides: carbide, boride, nitride, silicide
- Composite materials: particulate reinforced, fiber reinforced, combinations of oxides and nonoxides.

Each one of these classes can develop unique material properties because ceramics tend to be crystalline.

Applications

- Knife blades: the blade of a ceramic knife will stay sharp for much longer than that of a steel knife, although it is more brittle and can snap from a fall onto a hard surface.
- Carbon-ceramic brake disks for vehicles are resistant to brake fade at high temperatures.
- Advanced composite ceramic and metal matrices have been designed for most modern armoured fighting vehicles because they offer superior penetrating resistance against shaped charges (such as HEAT rounds) and kinetic energy penetrators.

- Ceramics such as alumina and boron carbide have been used in ballistic armored vests to repel large-caliber rifle fire. Such plates are known commonly as small arms protective inserts, or SAPIs. Similar material is used to protect the cockpits of some military airplanes, because of the low weight of the material.

- Ceramics can be used in place of steel for ball bearings. Their higher hardness means they are much less susceptible to wear and typically last for triple the lifetime of a steel part. They also deform less under load, meaning they have less contact with the bearing retainer walls and can roll faster. In very high speed applications, heat from friction during rolling can cause problems for metal bearings, which are reduced by the use of ceramics. Ceramics are also more chemically resistant and can be used in wet environments where steel bearings would rust. In some cases, their electricity-insulating properties may also be valuable in bearings. Two drawbacks to ceramic bearings are a significantly higher cost and susceptibility to damage under shock loads.

- In the early 1980s, Toyota researched production of an adiabatic engine using ceramic components in the hot gas area. The ceramics would have allowed temperatures of over 3000 °F (1650 °C). The expected advantages would have been lighter materials and a smaller cooling system (or no need for one at all), leading to a major weight reduction. The expected increase of fuel efficiency of the engine (caused by the higher temperature, as shown by Carnot's theorem) could not be verified experimentally; it was found that the heat transfer on the hot ceramic cylinder walls was higher than the transfer to a cooler metal wall as the cooler gas film on the metal surface works as a thermal insulator. Thus, despite all of these desirable properties, such engines have not succeeded in production because of costs for the ceramic components and the limited advantages. (Small imperfections in the ceramic material with its low fracture toughness lead to cracks, which can lead to potentially dangerous equipment failure.) Such engines are possible in laboratory settings, but mass production is not feasible with current technology.

- Work is being done in developing ceramic parts for gas turbine engines. Currently, even blades made of advanced metal alloys used in the engines' hot section require cooling and careful limiting of operating temperatures. Turbine engines made with ceramics could operate more efficiently, giving aircraft greater range and payload for a set amount of fuel.

- Recent advances have been made in ceramics which include bioceramics, such as dental implants and synthetic bones. Hydroxyapatite, the natural mineral component of bone, has been made synthetically from a number of biological and chemical sources and can be formed into ceramic materials. Orthopedic implants coated with these materials bond readily to bone and other tissues in the body without rejection or inflammatory reactions so are of great interest for gene delivery and tissue engineering scaffolds. Most hydroxyapatite ceramics are very porous and lack mechanical strength, and are used to coat metal orthopedic devices to aid in forming a bond to bone or as bone fillers. They are also used as fillers for orthopedic plastic screws to aid in reducing the inflammation and increase absorption of these plastic materials. Work is being done to make strong, fully dense nanocrystalline hydroxyapatite ceramic materials for orthopedic weight bearing devices, replacing foreign metal and plastic orthopedic materials with a synthetic, but naturally occurring, bone mineral. Ultimately, these ceramic materials may be used as bone replacements or with the incorporation of protein collagens, synthetic bones.

- High-tech ceramic is used in watchmaking for producing watch cases. The material is valued by watchmakers for its light weight, scratch resistance, durability and smooth touch. IWC is one of the brands that initiated the use of ceramic in watchmaking.

Ceramics in Archaeology

Ceramic artifacts have an important role in archaeology for understanding the culture, technology and behavior of peoples of the past. They are among the most common artifacts to be found at an archaeological site, generally in the form of small fragments of broken pottery called sherds. Processing of collected sherds can be consistent with two main types of analysis: technical and traditional.

Traditional analysis involves sorting ceramic artifacts, sherds and larger fragments into specific types based on style, composition, manufacturing and morphology. By creating these typologies it is possible to distinguish between different cultural styles, the purpose of the ceramic and technological state of the people among other conclusions. In addition, by looking at stylistic changes of ceramics over time is it possible to separate (seriate) the ceramics into distinct diagnostic groups (assemblages). A comparison of ceramic artifacts with known dated assemblages allows for a chronological assignment of these pieces.

The technical approach to ceramic analysis involves a finer examination of the composition of ceramic artifacts and sherds to determine the source of the material and through this the possible manufacturing site. Key criteria are the composition of the clay and the temper used in the manufacture of the article under study: temper is a material added to the clay during the initial production stage, and it is used to aid the subsequent drying process. Types of temper include shell pieces, granite fragments and ground sherd pieces called 'grog'. Temper is usually identified by microscopic examination of the temper material. Clay identification is determined by a process of refiring the ceramic, and assigning a color to it using Munsell Soil Color notation. By estimating both the clay and temper compositions, and locating a region where both are known to occur, an assignment of the material source can be made. From the source assignment of the artifact further investigations can be made into the site of manufacture.

Polymers

- Polymers – Chain of H-C molecules. Each repeat unit of H-C is a monomer e.g. ethylene (C_2H_4), Polyethylene – $(-CH_2 - CH_2)_n$
- Polymers: Thermosets – Soften when heated and harden on cooling – totally reversible. Thermoplasts – Do not soften on heating
- Plastics – moldable into many shape and have sufficient structural rigidity. Are one of the most commonly used class of materials.
- Are used in clothing, housing, automobiles, aircraft, packaging, electronics, signs, recreation items, and medical implants.
- Natural plastics – hellac, rubber, asphalt, and cellulose.

Characteristics and Applications of some common Thermoplastics

Material	Characteristics	Applications
Polyethylene	Chemically resistant, tough, low friction coeff., low strength	Flexible bottles, toys, battery parts, ice trays, film wrapping materials
Polyamide (Nylon)	Good strength and toughness, abrasion resistant, liquid absorber, low friction coeff.	Bearings, gears, cams, bushings and jacketing for wires and cables
Fluorocarbon (Teflon)	Chemically inert, excellent electrical properties, relatively weak	Anticorrosive seals, chemical pipes and valves, bearings, anti-adhesive coatings, high temp electronic parts
Polyester (PET)	Tough plastic film, excellent fatigue and tear strength, corrosion resistant	Recording tapes, clothing, automotive tyre cords, beverage containers
Vinyl	Low-cost general purpose material, rigid, can be made flexible	Floor coverings, pipe, electric al wire insulation, garden hose, phonograph records
Polystyrene	Excellent electrical prop and optical clarity, good thermal and dimensional stability	Wall tile, battery cases, toys, lighting panels, housing appliances

Characteristics and Applications of some common Thermosetting Polymers

Material	Characteristics	Applications
Epoxy (Araldite)	Excellent mechanical properties and corrosion resistance, good electrical prop., good adhesion and dimensional stability	Electrical moldings, sinks, adhesives, protective coatings, fiber reinforced plastic (FRP), laminates
Phenolic (Bakelite)	Excellent thermal stability (>150°C), inexpensive, can be compounded with many resins	Motor housings, telephones, auto distributors, electrical fixtures
Polyester (Aropol)	Excellent electrical properties, low cost, can formulated for room or high temperature, often fiber reinforced	Helmets, fiberglass boats, auto body components, chair fans

Elastomers

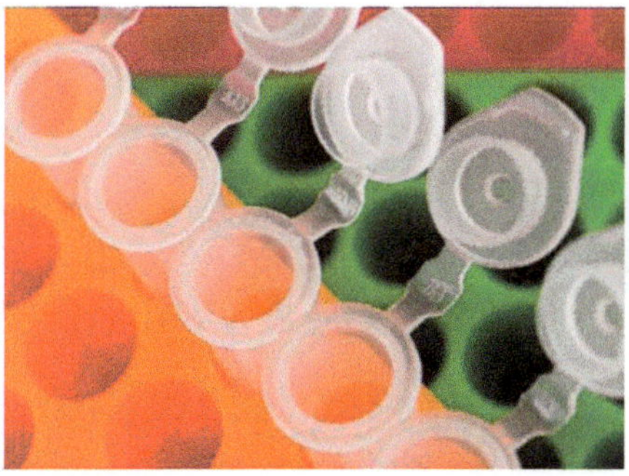

- Elastomer – a polymer with rubber-like elasticity.
- Each of the monomers that link to form the polymer is usually made of carbon, hydrogen, oxygen and/or silicon.
- Cross-linking in the monomers provides the flexibility.
- Glass transition temperature, T_g, is the temperature at which transition from rubbery to rigid state takes place in polymers.
- Elastomers are amorphous polymers existing above their T_g. Hence, considerable segmental motion exists in them.
- Their primary uses are in seals, adhesives and molded flexible parts.

Characteristics and Applications of some commercial Elastomers

Material	Characteristics	Applications
Natural rubber (NR)	Useful temp. range : - 60 – 120°Cgood resistance to cutting abrasion, resistant to oil, ozone, elng. 500- 700%	Pneumatic tyres and tubes, heels and soles, gaskets
Styrene-butadiene copolymer (SBR))	Temp. range: - 60 – 120° C, Good physical properties, elongation 450 – 500%	Same as natural rubber
Acrylonitrile-butadiene copolymer (NBR)	Temp. range: - 50 – 150° C. Excellent resistance to oils, elongation 400 – 600%	Gasoline, chemicals and oil hose, seals and O-rings, soles and heels

Liquid Crystal Polymers (LCP)

- LCPs are a group of chemically complex structure having unique properties. Primarily used in LCDs (liquid crystal displays) on watches, flat panel computer monitors, televisions and clocks.
- Advantages - LCDs are thinner and lighter and consume much less power than cathode ray tubes (CRTs).
- The name "liquid crystal" arises out of their characteristics. It takes a fair amount of heat to change a suitable substance from a solid into a liquid crystal, and it only takes a little more heat to turn the liquid crystal into a real liquid.
- LCDs use these liquid crystals because they react predictably to electric current in such a way as to control light passage.
- Aparticular sort of nematic liquid crystal, called twisted nematics (TN), is naturally twisted. Applying an electric current to these liquid crystals will untwist them to varying degrees, depending on the applied electrical potential.

Advanced Polymers

Ultrahigh Molecular Weight Polyethylene (UHMWPE)

- Molecular weight ca. 4×10^6 g/mol
- Outstanding properties
    - o high impact strength

- resistance to wear/abrasion
- low coefficient of friction
- self-lubricating surface

- Important applications
  - bullet-proof vests
  - golf ball covers
  - hip implants (acetabular cap)

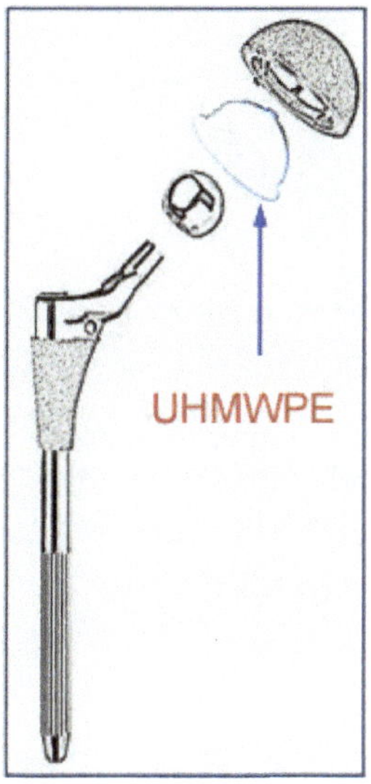

References

- Tunay, Olcay; Kabdasli, Isik; Arslan-Alaton, Idil; Olmez-Hanci, Tugba (2010-10-12). Chemical Oxidation Applications for Industrial Wastewaters. IWA Publishing. ISBN 9781843393078
- John C. Martin. "What we learn from a star's metal content". New Analysis RR Lyrae Kinematics in the Solar Neighborhood. Retrieved September 7, 2005
- Garvie, R. C.; Hannink, R. H.; Pascoe, R. T. (1975). "Ceramic steel?". Nature. 258 (5537): 703–704. Bibcode:1975Natur.258..703G. doi:10.1038/258703a0
- Abdul-Rahman, Yahia (2014-11-10). The Art of RF (Riba-Free) Islamic Banking and Finance: Tools and Techniques for Community-Based Banking. John Wiley & Sons. ISBN 9781118770962
- Mississippi Valley Archaeological Center, Ceramic Analysis Archived June 3, 2012, at the Wayback Machine., Retrieved 04-11-12
- Wachtman, John B., Jr. (ed.) (1999) Ceramic Innovations in the 20th century, The American Ceramic Society. ISBN 978-1-57498-093-6

3

Phase Transformation in Materials

The state of matter when the physical properties of any material remain constant is called phase. The physical properties include density, magnetization, index of refraction, chemical composition etc. The phases of a material are studied using a phase diagram, which is a graph that charts the phases in relation to its pressure, temperature and volume. The major components of phase transformation in materials are discussed in this chapter.

Phase (Matter)

In the physical sciences, a phase is a region of space (a thermodynamic system), throughout which all physical properties of a material are essentially uniform. Examples of physical properties include density, index of refraction, magnetization and chemical composition. A simple description is that a phase is a region of material that is chemically uniform, physically distinct, and (often) mechanically separable. In a system consisting of ice and water in a glass jar, the ice cubes are one phase, the water is a second phase, and the humid air over the water is a third phase. The glass of the jar is another separate phase.

A small piece of rapidly melting argon ice shows the transition from solid to liquid.

The term *phase* is sometimes used as a synonym for state of matter, but there can be several immiscible phases of the same state of matter. Also, the term *phase* is sometimes used to refer to a set of equilibrium states demarcated in terms of state variables such as pressure and temperature by a phase boundary on a phase diagram. Because phase boundaries relate to changes in the organization of matter, such as a change from liquid to solid or a more subtle change from one crystal structure to another, this latter usage is similar to the use of "phase" as a synonym for state of matter. However, the state of matter and phase diagram usages are not commensurate with the formal definition given above and the intended meaning must be determined in part from the context in which the term is used.

Types of Phases

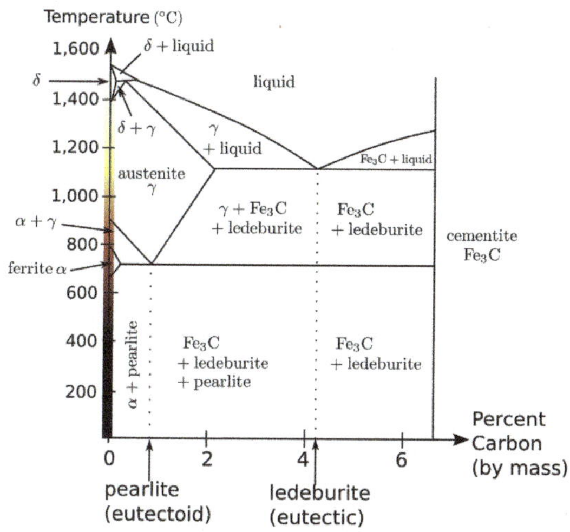

Iron-carbon phase diagram, showing the conditions necessary to form different phases

Distinct phases may be described as different states of matter such as gas, liquid, solid, plasma or Bose–Einstein condensate. Useful mesophases between solid and liquid form other states of matter.

Distinct phases may also exist within a given state of matter. As shown in the diagram for iron alloys, several phases exist for both the solid and liquid states. Phases may also be differentiated based on solubility as in polar (hydrophilic) or non-polar (hydrophobic). A mixture of water (a polar liquid) and oil (a non-polar liquid) will spontaneously separate into two phases. Water has a very low solubility (is insoluble) in oil, and oil has a low solubility in water. Solubility is the maximum amount of a solute that can dissolve in a solvent before the solute ceases to dissolve and remains in a separate phase. A mixture can separate into more than two liquid phases and the concept of phase separation extends to solids, i.e., solids can form solid solutions or crystallize into distinct crystal phases. Metal pairs that are mutually soluble can form alloys, whereas metal pairs that are mutually insoluble cannot.

As many as eight immiscible liquid phases have been observed. Mutually immiscible liquid phases are formed from water (aqueous phase), hydrophobic organic solvents, perfluorocarbons (fluorous phase), silicones, several different metals, and also from molten phosphorus. Not all organic solvents are completely miscible, e.g. a mixture of ethylene glycol and toluene may separate into two distinct organic phases.

Phases do not need to macroscopically separate spontaneously. Emulsions and colloids are examples of immiscible phase pair combinations that do not physically separate.

Phase Equilibrium

Left to equilibration, many compositions will form a uniform single phase, but depending on the temperature and pressure even a single substance may separate into two or more distinct phases. Within each phase, the properties are uniform but between the two phases properties differ.

Water in a closed jar with an air space over it forms a two phase system. Most of the water is in the liquid phase, where it is held by the mutual attraction of water molecules. Even at equilibrium molecules are constantly in motion and, once in a while, a molecule in the liquid phase gains enough kinetic energy to break away from the liquid phase and enter the gas phase. Likewise, every once in a while a vapor molecule collides with the liquid surface and condenses into the liquid. At equilibrium, evaporation and condensation processes exactly balance and there is no net change in the volume of either phase.

At room temperature and pressure, the water jar reaches equilibrium when the air over the water has a humidity of about 3%. This percentage increases as the temperature goes up. At 100 °C and atmospheric pressure, equilibrium is not reached until the air is 100% water. If the liquid is heated a little over 100 °C, the transition from liquid to gas will occur not only at the surface, but throughout the liquid volume: the water boils.

Number of Phases

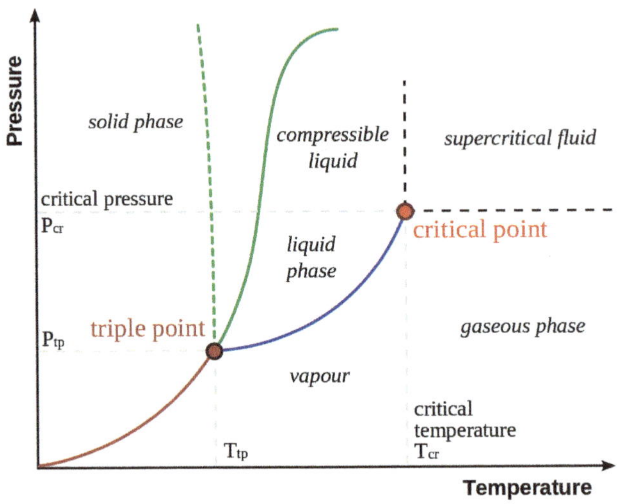

A typical phase diagram for a single-component material, exhibiting solid, liquid and gaseous phases. The solid green line shows the usual shape of the liquid–solid phase line. The dotted green line shows the anomalous behavior of water when the pressure increases. The triple point and the critical point are shown as red dots.

For a given composition, only certain phases are possible at a given temperature and pressure. The number and type of phases that will form is hard to predict and is usually determined by experiment. The results of such experiments can be plotted in phase diagrams.

The phase diagram shown here is for a single component system. In this simple system, which phases that are possible depends only on pressure and temperature. The markings show points where two or more phases can co-exist in equilibrium. At temperatures and pressures away from the markings, there will be only one phase at equilibrium.

In the diagram, the blue line marking the boundary between liquid and gas does not continue indefinitely, but terminates at a point called the critical point. As the temperature and pressure approach the critical point, the properties of the liquid and gas become progressively more similar. At the critical point, the liquid and gas become indistinguishable. Above the critical point, there are no longer separate liquid and gas phases: there is only a generic fluid phase referred to as a supercritical fluid. In water, the critical point occurs at around 647 K (374 °C or 705 °F) and 22.064 MPa.

An unusual feature of the water phase diagram is that the solid–liquid phase line (illustrated by the dotted green line) has a negative slope. For most substances, the slope is positive as exemplified by the dark green line. This unusual feature of water is related to ice having a lower density than liquid water. Increasing the pressure drives the water into the higher density phase, which causes melting.

Another interesting though not unusual feature of the phase diagram is the point where the solid–liquid phase line meets the liquid–gas phase line. The intersection is referred to as the triple point. At the triple point, all three phases can coexist.

Experimentally, the phase lines are relatively easy to map due to the interdependence of temperature and pressure that develops when multiple phases forms. Consider a test apparatus consisting of a closed and well insulated cylinder equipped with a piston. By charging the right amount of water and applying heat, the system can be brought to any point in the gas region of the phase diagram. If the piston is slowly lowered, the system will trace a curve of increasing temperature and pressure within the gas region of the phase diagram. At the point where gas begins to condense to liquid, the direction of the temperature and pressure curve will abruptly change to trace along the phase line until all of the water has condensed.

Interfacial Phenomena

Between two phases in equilibrium there is a narrow region where the properties are not that of either phase. Although this region may be very thin, it can have significant and easily observable effects, such as causing a liquid to exhibit surface tension. In mixtures, some components may preferentially move toward the interface. In terms of modeling, describing, or understanding the behavior of a particular system, it may be efficacious to treat the interfacial region as a separate phase.

Crystal Phases

A single material may have several distinct solid states capable of forming separate phases. Water is a well-known example of such a material. For example, water ice is ordinarily found in the hexagonal form ice I_h, but can also exist as the cubic ice I_c, the rhombohedral ice II, and many other forms. Polymorphism is the ability of a solid to exist in more than one crystal form. For pure chemical elements, polymorphism is known as allotropy. For example, diamond, graphite, and fullerenes are different allotropes of carbon.

Phase Transitions

When a substance undergoes a phase transition (changes from one state of matter to another) it usually either takes up or releases energy. For example, when water evaporates, the increase in kinetic energy as the evaporating molecules escape the attractive forces of the liquid is reflected in a decrease in temperature. The energy required to induce the phase transition is taken from the internal thermal energy of the water, which cools the liquid to a lower temperature; hence evaporation is useful for cooling. The reverse process, condensation, releases heat. The heat energy, or enthalpy, associated with a solid to liquid transition is the enthalpy of fusion and that associated with a solid to gas transition is the enthalpy of sublimation.

Phase Diagram

A phase diagram in physical chemistry, engineering, mineralogy, and materials science is a type of chart used to show conditions (pressure, temperature, volume, etc.) at which thermodynamically distinct phases occur and coexist at equilibrium.

Overview

Common components of a phase diagram are *lines of equilibrium* or *phase boundaries*, which refer to lines that mark conditions under which multiple phases can coexist at equilibrium. Phase transitions occur along lines of equilibrium.

Triple points are points on phase diagrams where lines of equilibrium intersect. Triple points mark conditions at which three different phases can coexist. For example, the water phase diagram has a triple point corresponding to the single temperature and pressure at which solid, liquid, and gaseous water can coexist in a stable equilibrium (273.16 K and a partial vapor pressure of 611.657 Pa).

The solidus is the temperature below which the substance is stable in the solid state. The liquidus is the temperature above which the substance is stable in a liquid state. There may be a gap between the solidus and liquidus; within the gap, the substance consists of a mixture of crystals and liquid (like a "slurry").

Types

2-dimensional Diagrams

The simplest phase diagrams are pressure–temperature diagrams of a single simple substance, such as water. The axes correspond to the pressure and temperature. The phase diagram shows, in pressure–temperature space, the lines of equilibrium or phase boundaries between the three phases of solid, liquid, and gas.

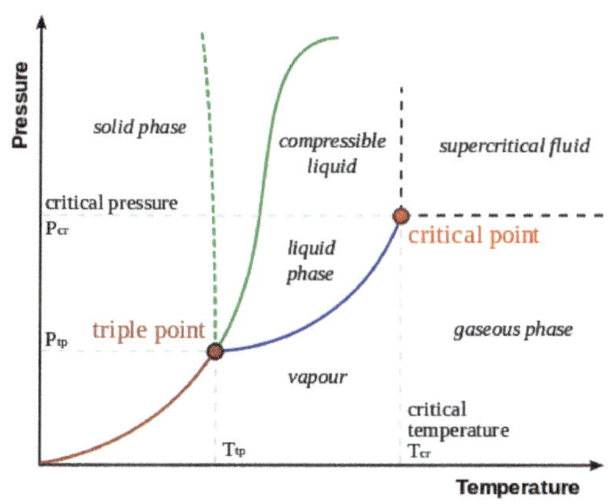

A typical phase diagram. The solid green line applies to most substances; the dotted green line gives the anomalous behavior of water. The green lines mark the freezing point and the blue line the boiling point, showing how they vary with pressure.

The fusion curves for water and Antimony are with negative slopes, All other liquids have positive slope on P-T diagram.

The curves on the phase diagram show the points where the free energy (and other derived properties) becomes non-analytic: their derivatives with respect to the coordinates (temperature and pressure in this example) change discontinuously (abruptly). For example, the heat capacity of a container filled with ice will change abruptly as the container is heated past the melting point. The open spaces, where the free energy is analytic, correspond to single phase regions. Single phase regions are separated by lines of non-analytical behavior, where phase transitions occur, which are called phase boundaries.

In the diagram on the left, the phase boundary between liquid and gas does not continue indefinitely. Instead, it terminates at a point on the phase diagram called the critical point. This reflects the fact that, at extremely high temperatures and pressures, the liquid and gaseous phases become indistinguishable, in what is known as a supercritical fluid. In water, the critical point occurs at around T_c = 647.096 K (373.946 °C), p_c = 22.064 MPa (217.75 atm) and ρ_c = 356 kg/m³.

The existence of the liquid–gas critical point reveals a slight ambiguity in labelling the single phase regions. When going from the liquid to the gaseous phase, one usually crosses the phase boundary, but it is possible to choose a path that never crosses the boundary by going to the right of the critical point. Thus, the liquid and gaseous phases can blend continuously into each other. The solid–liquid phase boundary can only end in a critical point if the solid and liquid phases have the same symmetry group.

The solid–liquid phase boundary in the phase diagram of most substances has a positive slope; the greater the pressure on a given substance, the closer together the molecules of the substance are brought to each other, which increases the effect of the substance's intermolecular forces. Thus, the substance requires a higher temperature for its molecules to have enough energy to break out of the fixed pattern of the solid phase and enter the liquid phase. A similar concept applies to liquid–gas phase changes. Water, because of its particular properties, is one of the several exceptions to the rule.

Other Thermodynamic Properties

In addition to temperature and pressure, other thermodynamic properties may be graphed in phase diagrams. Examples of such thermodynamic properties include specific volume, specific enthalpy, or specific entropy. For example, single-component graphs of temperature vs. specific entropy (T vs. s) for water/steam or for a refrigerant are commonly used to illustrate thermodynamic cycles such as a Carnot cycle, Rankine cycle, or vapor-compression refrigeration cycle.

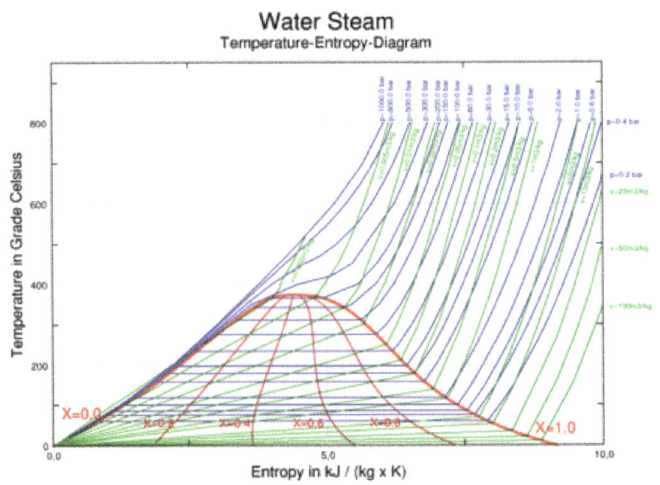

Temperature vs. specific entropy phase diagram for water/steam. In the area under the red dome, liquid water and steam coexist in equilibrium. The critical point is at the top of the dome. Liquid water is to the left of the dome. Steam is to the right of the dome. The blue lines/curves are isobars showing constant pressure. The green lines/curves are isochors showing constant specific volume. The red curves show constant quality.

In a two-dimensional graph, two of the thermodynamic quantities may be shown on the horizontal and vertical axes. Additional thermodynamic quantities may each be illustrated in increments as a series of lines - curved, straight, or a combination of curved and straight. Each of these iso-lines represents the thermodynamic quantity at a certain constant value.

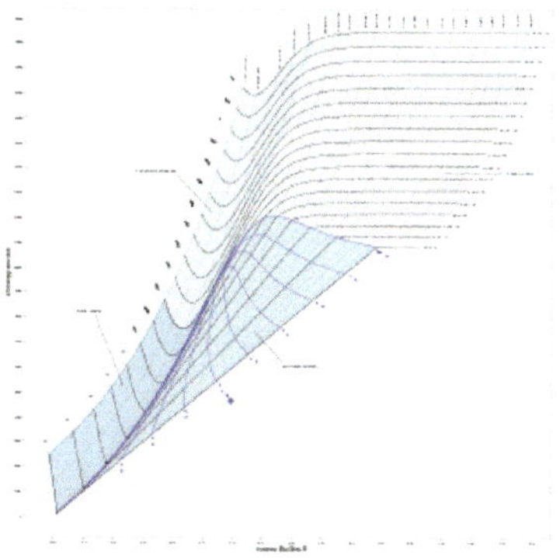

enthalpy–entropy (h–s) diagram for steam

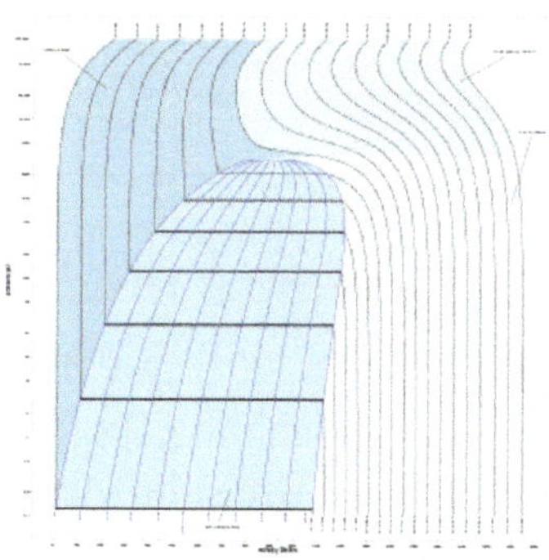

pressure–enthalpy (p–h) diagram for steam

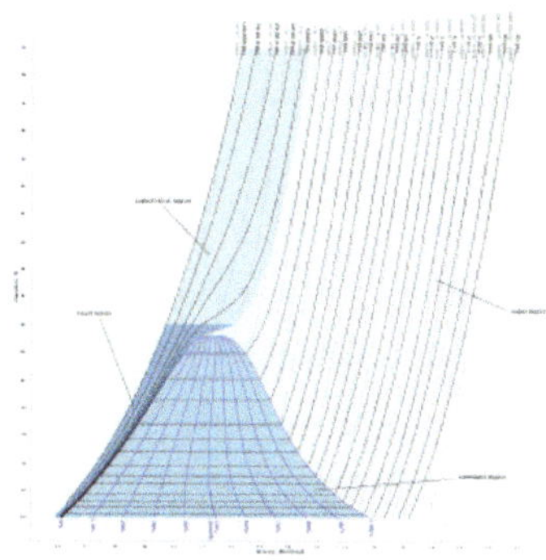

temperature–entropy (T–s) diagram for steam

3-dimensional Diagrams

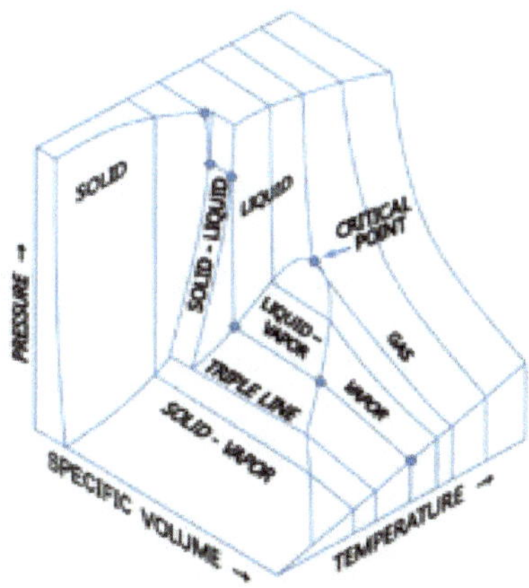

p–v–T 3D diagram for fixed amount of pure material

It is possible to envision three-dimensional (3D) graphs showing three thermodynamic quantities. For example, for a single component, a 3D Cartesian coordinate type graph can show temperature (T) on one axis, pressure (p) on a second axis, and specific volume (v) on a third. Such a 3D graph is sometimes called a p–v–T diagram. The equilibrium conditions are shown as curves on a curved surface in 3D with areas for solid, liquid, and vapor phases and areas where solid and liquid, solid and vapor, or liquid and vapor coexist in equilibrium. A line on the surface called a triple line is where solid, liquid and vapor can all coexist in equilibrium. The critical point remains a point on the surface even on a 3D phase diagram.

For water, the 3D $p–v–T$ diagram is seen here:

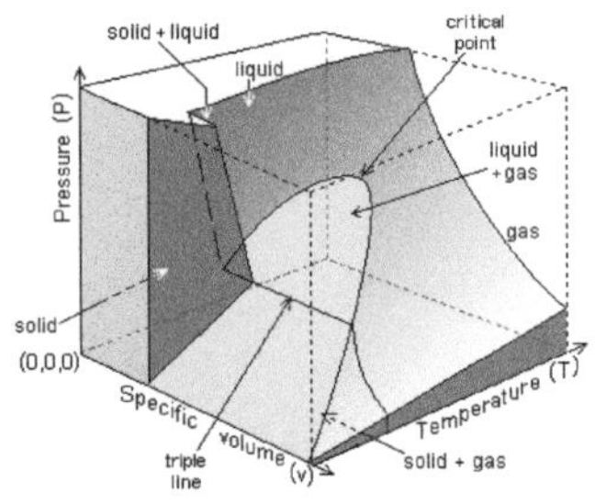

3D phase diagram of water fluids and selected ices

An orthographic projection of the 3D $p–v–T$ graph showing pressure and temperature as the vertical and horizontal axes collapses the 3D plot into the standard 2D pressure–temperature diagram. When this is done, the solid–vapor, solid–liquid, and liquid–vapor surfaces collapse into three corresponding curved lines meeting at the triple point, which is the collapsed orthographic projection of the triple line.

Binary Mixtures

Other much more complex types of phase diagrams can be constructed, particularly when more than one pure component is present. In that case, concentration becomes an important variable. Phase diagrams with more than two dimensions can be constructed that show the effect of more than two variables on the phase of a substance. Phase diagrams can use other variables in addition to or in place of temperature, pressure and composition, for example the strength of an applied electrical or magnetic field, and they can also involve substances that take on more than just three states of matter.

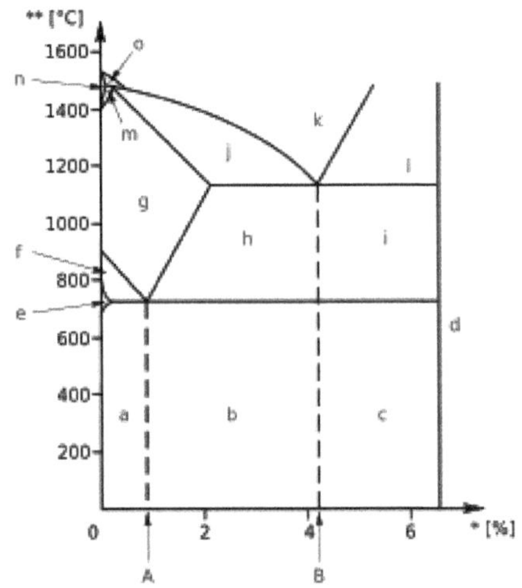

The iron–iron carbide ($Fe–Fe_3C$) phase diagram. The percentage of carbon present and the temperature define the phase of the iron carbon alloy and therefore its physical characteristics and mechanical properties. The percentage of carbon determines the type of the ferrous alloy: iron, steel or cast iron

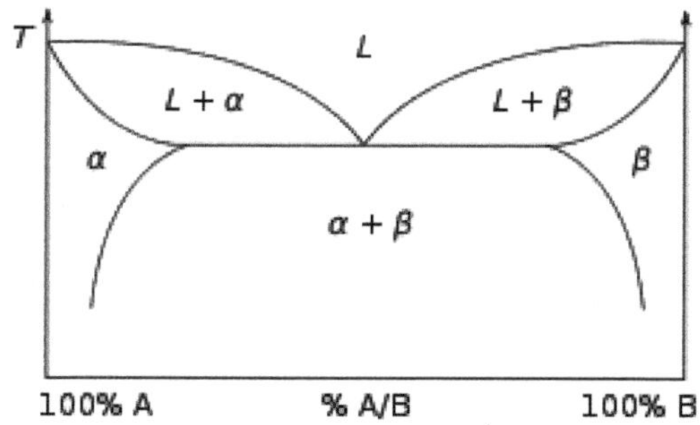

A phase diagram for a binary system displaying a eutectic point.

One type of phase diagram plots temperature against the relative concentrations of two substances in a binary mixture called a *binary phase diagram*, as shown at right. Such a mixture can be either a solid solution, eutectic or peritectic, among others. These two types of mixtures result in very different graphs. Another type of binary phase diagram is a *boiling-point diagram* for a mixture of two components, i. e. chemical compounds. For two particular volatile components at a certain pressure such as atmospheric pressure, a boiling-point diagram shows what vapor (gas) compositions are in equilibrium with given liquid compositions depending on temperature. In a typical binary boiling-point diagram, temperature is plotted on a vertical axis and mixture composition on a horizontal axis.

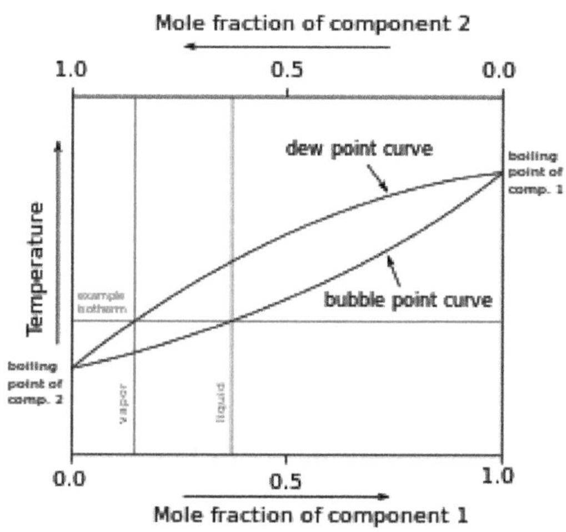

Boiling-point diagram

A simple example diagram with hypothetical components 1 and 2 in a non-azeotropic mixture is shown at right. The fact that there are two separate curved lines joining the boiling points of the pure components means that the vapor composition is usually not the same as the liquid composition the vapor is in equilibrium with.

In addition to the above-mentioned types of phase diagrams, there are thousands of other possible combinations. Some of the major features of phase diagrams include congruent points, where a solid phase transforms directly into a liquid. There is also the peritectoid, a point where two solid phases combine into one solid phase during cooling. The inverse of this, when one solid phase transforms into two solid phases during cooling, is called the eutectoid.

A complex phase diagram of great technological importance is that of the iron–carbon system for less than 7% carbon.

The x-axis of such a diagram represents the concentration variable of the mixture. As the mixtures are typically far from dilute and their density as a function of temperature is usually unknown, the preferred concentration measure is mole fraction. A volume-based measure like molarity would be inadvisable.

Crystals

Polymorphic and polyamorphic substances have multiple crystal or amorphous phases, which can be graphed in a similar fashion to solid, liquid, and gas phases.

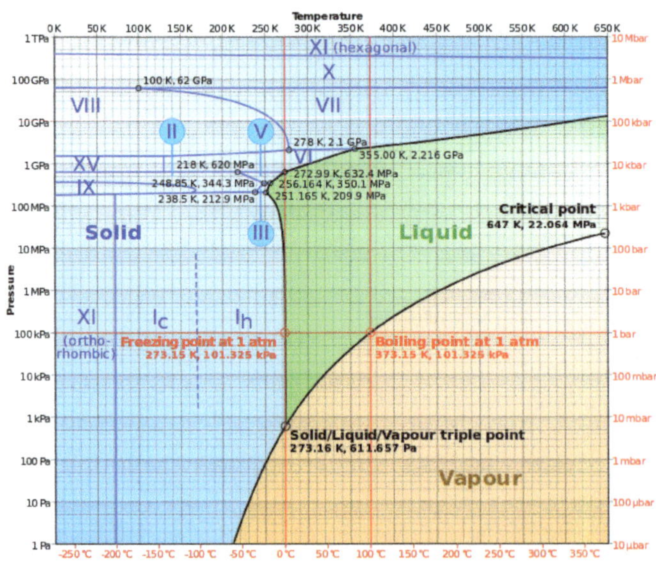

Log-lin pressure–temperature phase diagram of water. The Roman numerals indicate various ice phases. A similar diagram may be found on the site Water structure and science.

Mesophases

Some organic materials pass through intermediate states between solid and liquid; these states are called mesophases. Attention has been directed to mesophases because they enable display devices and have become commercially important through the so-called liquid-crystal technology. Phase diagrams are used to describe the occurrence of mesophases.

- The composition of phases in the two-phase region is not same.
- To find the composition of the individual phases in the two- phase region, a hori-

zontal line (XY), called *tie line*, is drawn and its intercepts on the liquidus and solidus lines, C_l and C_s, are taken as the composition of the liquid and solid respectively

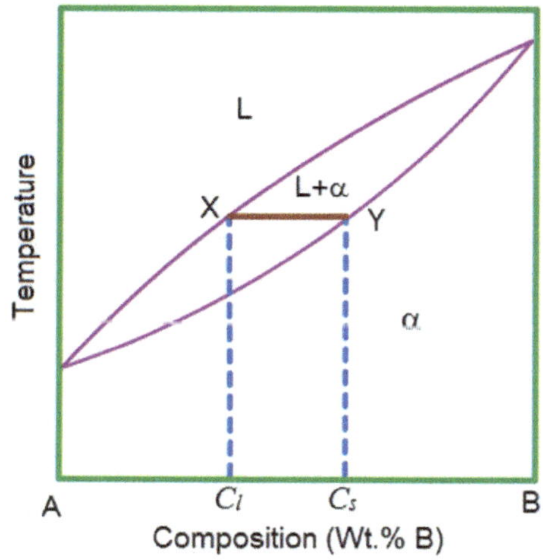

Lever Rule

- The relative fractions of the phases at a given temperature for an alloy composition C_0 is obtained by the *lever rule*. This rule gives the fraction of a phase by the ratio of the lengths of the tie line between C_0 and composition of the other phase to the total length of the tie line. For example, fraction solid, f_S is given by

$$f_s = \frac{MX}{XY} = \frac{C_o - C_l}{C_s - C_l}$$

Similarly fraction liquid, f_l

$$f_l = \frac{MY}{XY} = \frac{C_s - C_o}{C_s - C_l}$$

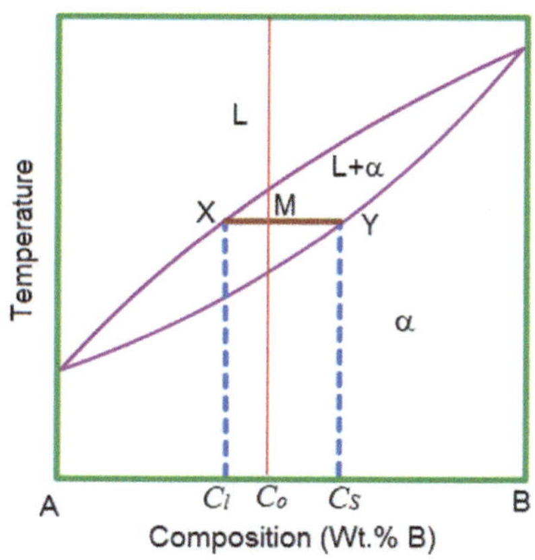

Cooling Curves

- Upon cooling from liquid state, the temperature of the pure metal (A or B) drops continuously till melting point at which solidification starts. Solidification happens at a constant temperature (line PQ) as F =0 (F = 1 – 2 +1 = 0). The temperature drops again on completion of solidification.
- For any alloy (1, 2, 3 etc.) temp. drops till the liquidus (L_1, L_2, L_3). However, in this case, solidification proceeds over a range of temperature as F = 1 (2 – 2 + 1 = 1). Once solidification completes at the solidus (S_1, S_2, S_3) the temp. drops again.

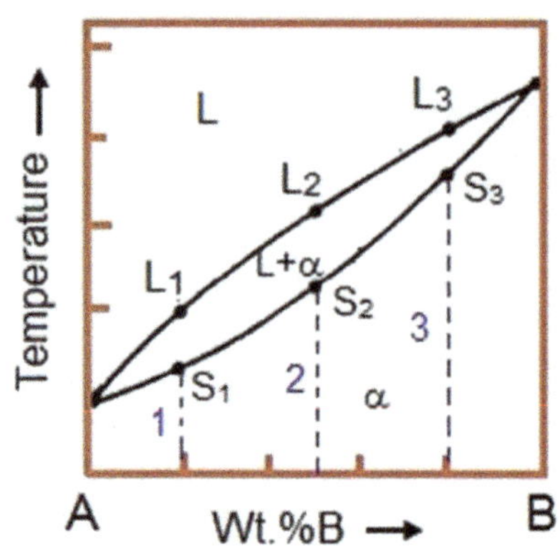

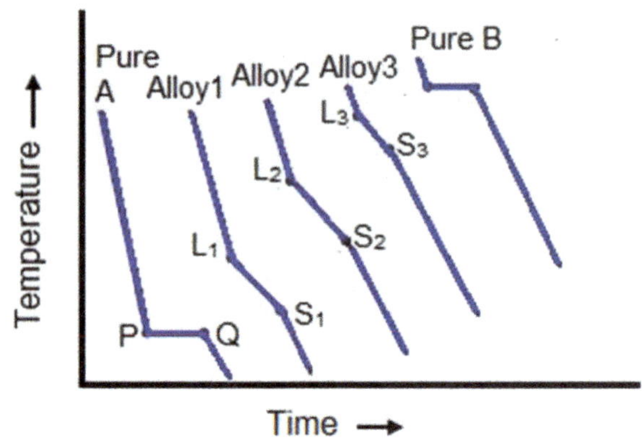

Phase Diagrams- Limited Solubility

- Not all metals are completely soluble in each other. Distinctions can be made between two types solid solutions with limited solubility – (i) Eutectic and (ii) Peritectic.
- When the melting points of two metals are comparable, a eutectic system forms while a peritectic results when melting points are significantly different.
- A eutectic reaction is defined as the one which generates two solids from the liquid at a given temperature and composition, $L \rightarrow \alpha + \beta$
- Peritectic is Liquid + Solid 1 → Solid 2 ($L + \alpha \rightarrow \beta$)
- In both the cases three phases (two solids and a liquid) coexist and the degrees of freedom F = 2 – 3 + 1 = 0. This is known as invariant (F = 0) reaction or transformation.

Eutectic Phase diagram

- In the eutectic system between two metals A and B, two solid solutions, one rich in A (α) and another rich in B (β) form.

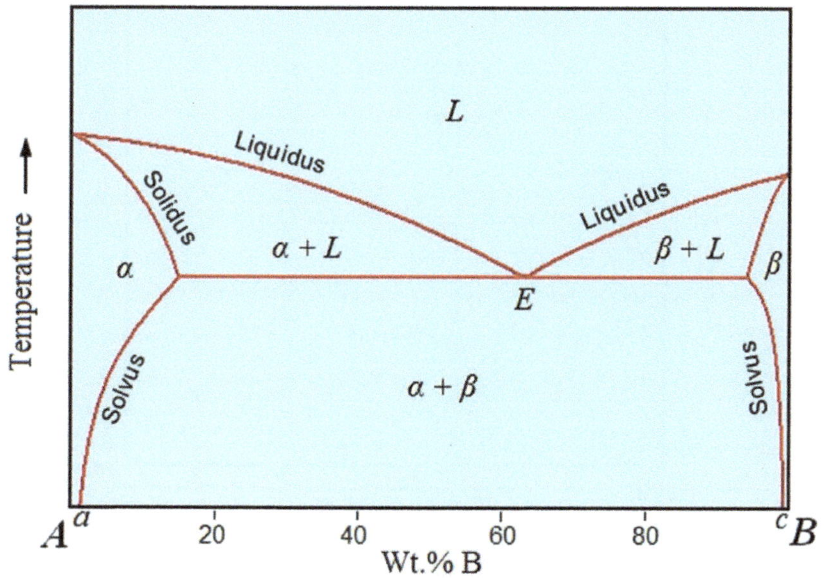

- In addition to liquidus and solidus lines there are two more lines on A and B rich ends which define the solubility limits B in A and A in B respectively. These are called solvus lines.

- Three phases ($L+\alpha+\beta$) coexist at point *E*. This point is called eutectic point or composition. Left of *E* is called hypoeutectic whereas right of E is called hypereutectic.

- A eutectic composition solidifies as a eutectic mixture of α and β phases. The microstructure at room temperature (RT) may consist of alternate layers or lamellae of α and β.

- In hypoeutectic alloys the α phase solidifies first and the microstructure at RT consists of this α phase (called proeutectic α) and the eutectic ($\alpha+\beta$) mixture. Similarly hypereutectic alloys consist of proeutectic β and the eutectic mixture.

- The melting point at the eutectic point is minimum. That's why Pb-Sn eutectic alloys are used as solders. Other eutectic systems are Ag-Cu, Al-Si, Al-Cu.

Eutectic Cooling Curves

- While cooling a hypoeutectic alloy from the liquid state, the temp. drops continuously till liquidus point, *a*, at which crystals of proeutectic α begins to form.

- On further cooling the fraction of α increases. At any point, *b*, in the two-phase region the α fraction is given by the *lever rule* as *bn/mn*.

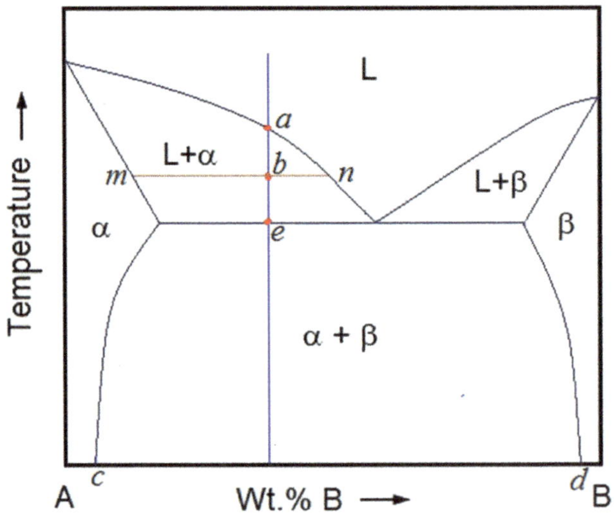

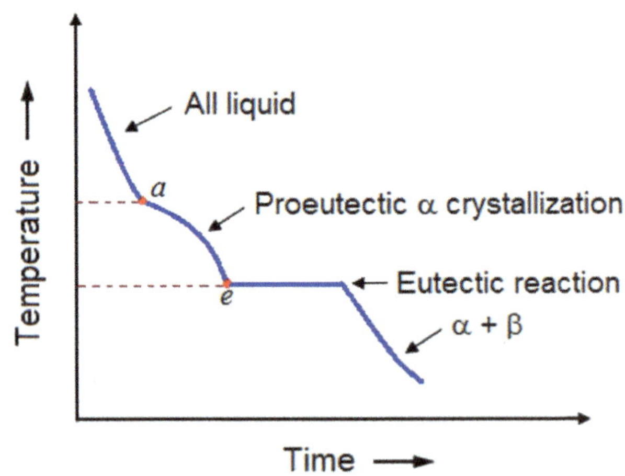

- Solidification of proeutectic α continues till the eutectic temperature is reached. The inflection in the cooling curve between points *a* and *e* is due to evolution of the latent heat.
- At the eutectic point (*e*) the solidification of eutectic mixture $(\alpha + \beta)$ begins through the eutectic reaction and proceeds at a constant temperature as F = 0 (2 – 3 + 1).
- The cooling behavior in hypereutectic alloy is similar except that proeutectic β forms below the liquidus.
- For a eutectic composition, the proeutectic portion is absent and the cooling curve appears like that of a pure metal.
- Any composition left of point *c* or right of point *d* (α and β single phase region respectively) will cool and solidify like an isomorphous system.

Peritectic Phase Diagram

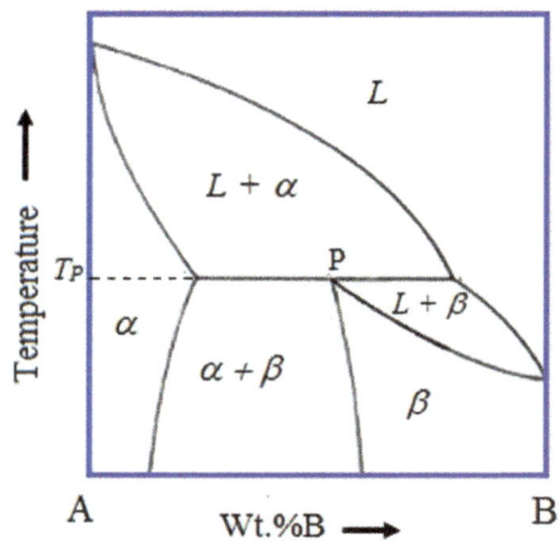

- L + α → β. An alloy cooling slowly through the peritectic point, *P*, the α phase will crystallize first just below the liquidus line. At the peritectic temperature, T_P all of the liquid and α will convert to β .
- Any composition left of *P* will generate excess α and similarly compositions right of *P* will give rise to an excess of liquid.
- Peritectic systems – Pt - Ag, Ni - Re, Fe - Ge, Sn-Sb (babbit).

Monotectic Phase Diagram

- Another three phase invariant reaction that occurs in some binary system is monotectic reaction in which a liquid transforms to another liquid and a solid. $L_1 \rightarrow L_2 + \alpha$.
- Two liquids are immiscible like water and oil over certain range of compositions. Cu-Pb system has a monotectic at 36% Pb and 955 °C.

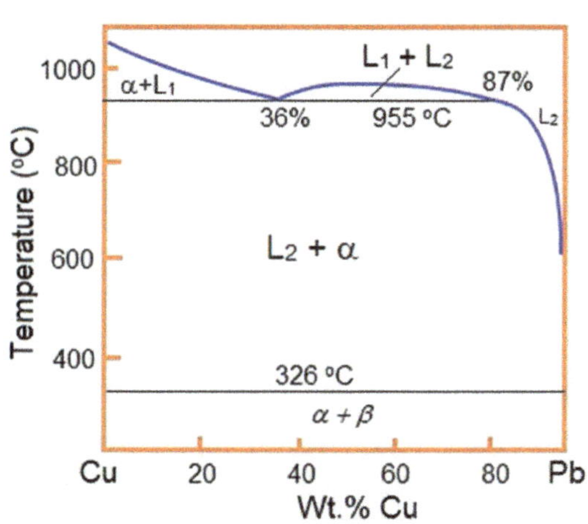

Cu-Pd system – Monotectic portion

Phase Diagrams with Intermediate Phases

- Binary system can have two types of solid solutions/phases terminal phases and intermediate phases.
- Terminal phases occur near the pure metal ends, e.g. α and β phases in the eutectic system.
- Intermediate phases occur inside the phase diagram and are separated by two-phase regions.
- The Cu-Zn system contains both types of phases. α and η are terminal phases and β, γ, δ and ε are intermediate phases.
- Intermediate phases form in ceramic phase diagrams also. For example, in the $Al_2O_3 - SiO_2$ system an intermediate phase called mullite ($3Al_2O_3 - 2SiO_2$) is formed.

Intermediate Phases - Cu-Zn Phase Diagram

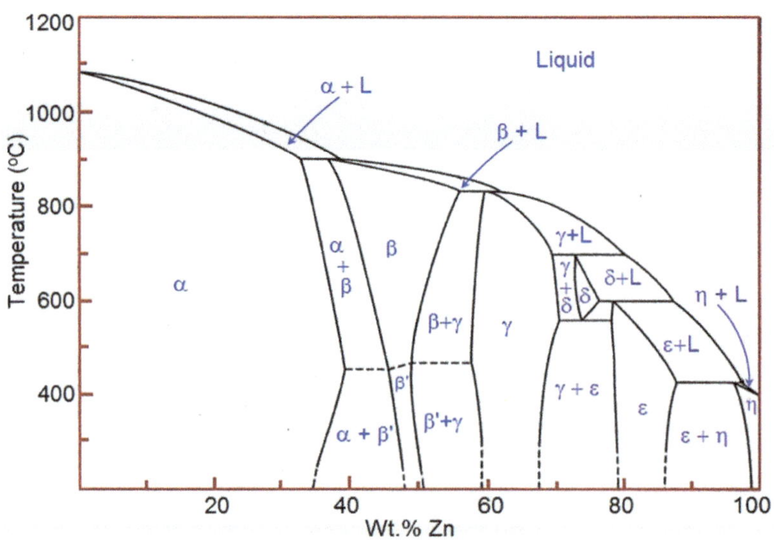

Cu-Zn phase diagram. α and η are terminal phases and β, γ, δ and ε are intermediate phases.

Phase Diagrams with Compounds

- Sometimes a crystalline compound called *intermetallic compound* may form between two metals.
- Such compounds generally have a distinct chemical formula or stoichiometry.
- Example – Mg_2Pb in the Mg-Pb system (appear as a vertical line at 81% Pb), Mg_2Ni, Mg_2Si, Fe_3C.

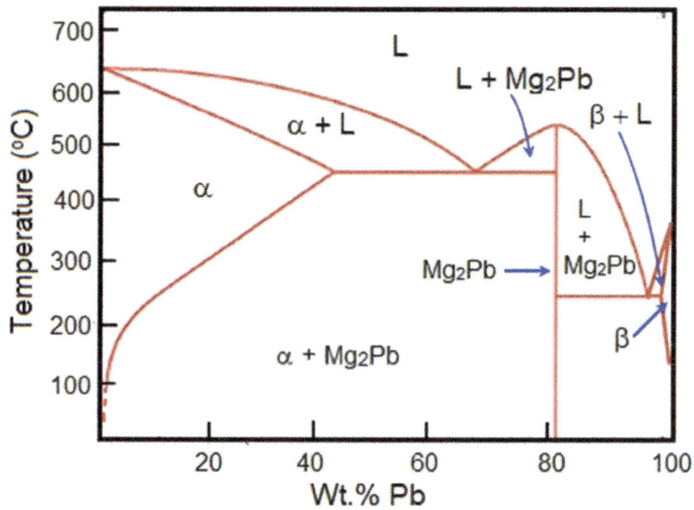

Ternary Phase Diagram

- A ternary or three component phase diagram has the form of an triangular prism with an equilateral triangle as a base.
- Pure components are at each vertex, sides are binary compositions and ternary compositions are within the triangle.
- The composition lines on the triangle is constructed from projections of surfaces.

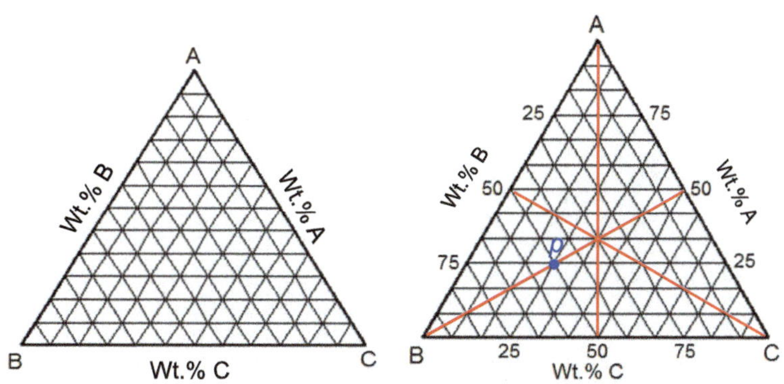

- The temperature varies along the height of the prism. The composition triangle is an isothermal section. Alternatively projections of different surfaces and lines can be shown as temperature contours.
- The composition of any point in the triangle is determined by drawing perpendiculars from corners to the opposite sides and measuring the distance of the point along the perpendicular.
- Point *p*, for example, lies on the isocomoposition line 25% A along the perpen-

dicular A-50. Hence, percentage of A in the alloy is 25%. Similarly B is 50% and C is 25%.

Phase Boundary

In thermal equilibrium, each phase (i.e. liquid, solid etc.) of physical matter comes to an end at a transitional point, or spatial interface, called a phase boundary, due to the immiscibility of said matter with the matter on the other side of said boundary. This immiscibility is due to at least one difference between the two substances' corresponding physical properties. The behavior of phase boundaries has been a developing subject of interest and an active research field, called interface science, in physics and mathematics for almost two centuries, due partly to phase boundaries naturally arising in many physical processes, such as the capillarity effect, the growth of grain boundaries, the physics of binary alloys, and the formation of snow flakes.

One of the oldest problems in the area dates back to Lamé and Clapeyron who studied the freezing of the ground. Their goal was to determine the thickness of solid crust generated by the cooling of a liquid at constant temperature filling the half-space. In 1889, Stefan, while working on the freezing of the ground developed these ideas further and formulated the two phase model which came to be known as the Stefan Problem.

The proof of existence and uniqueness of a solution to the Stefan problem was done in many stages. Proving the general existence and uniqueness of the solutions in the case of $d = 3$ was solved by Shoshana Kamin.

Phase Transition

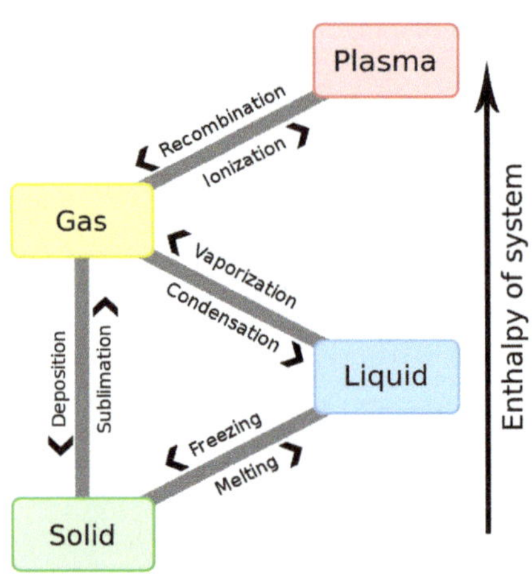

This diagram shows the nomenclature for the different phase transitions.

The term phase transition is most commonly used to describe transitions between solid, liquid and gaseous states of matter, and, in rare cases, plasma. A phase of a thermodynamic system and the states of matter have uniform physical properties. During a phase transition of a given medium certain properties of the medium change, often discontinuously, as a result of the change of some external condition, such as temperature, pressure, or others. For example, a liquid may become gas upon heating to the boiling point, resulting in an abrupt change in volume. The measurement of the external conditions at which the transformation occurs is termed the phase transition. Phase transitions are common in nature and used today in many technologies.

Types of Phase Transition

Examples of phase transitions include:

- The transitions between the solid, liquid, and gaseous phases of a single component, due to the effects of temperature and/or pressure:

		To			
		Solid	**Liquid**	**Gas**	**Plasma**
From	**Solid**	Solid-solid transformation	Melting	Sublimation	—
	Liquid	Freezing	—	Boiling/evaporation	—
	Gas	Deposition	Condensation	—	Ionization
	Plasma	—	—	Recombination/ deionization	—

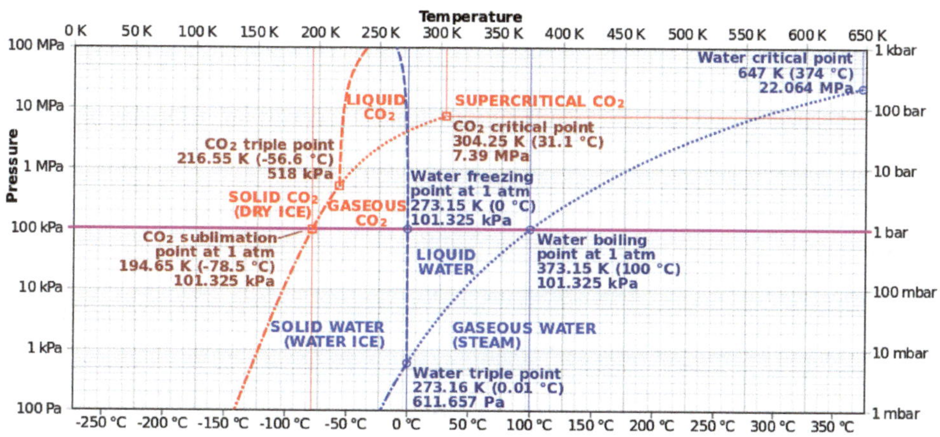

Comparison of phase diagrams of carbon dioxide (red) and water (blue) explaining their different phase transitions at 1 atmosphere

- A eutectic transformation, in which a two component single phase liquid is cooled and transforms into two solid phases. The same process, but beginning with a solid instead of a liquid is called a eutectoid transformation.

- A peritectic transformation, in which a two component single phase solid is heated and transforms into a solid phase and a liquid phase.
- A spinodal decomposition, in which a single phase is cooled and separates into two different compositions of that same phase.
- Transition to a mesophase between solid and liquid, such as one of the "liquid crystal" phases.
- The transition between the ferromagnetic and paramagnetic phases of magnetic materials at the Curie point.
- The transition between differently ordered, commensurate or incommensurate, magnetic structures, such as in cerium antimonide.
- The martensitic transformation which occurs as one of the many phase transformations in carbon steel and stands as a model for displacive phase transformations.
- Changes in the crystallographic structure such as between ferrite and austenite of iron.
- Order-disorder transitions such as in alpha-titanium aluminides.
- The dependence of the adsorption geometry on coverage and temperature, such as for hydrogen on iron (110).
- The emergence of superconductivity in certain metals and ceramics when cooled below a critical temperature.
- The transition between different molecular structures (polymorphs, allotropes or polyamorphs), especially of solids, such as between an amorphous structure and a crystal structure, between two different crystal structures, or between two amorphous structures.
- Quantum condensation of bosonic fluids (Bose–Einstein condensation). The superfluid transition in liquid helium is an example of this.
- The breaking of symmetries in the laws of physics during the early history of the universe as its temperature cooled.
- Isotope fractionation occurs during a phase transition, the ratio of light to heavy isotopes in the involved molecules changes. When water vapor condenses (an equilibrium fractionation), the heavier water isotopes (18O and 2H) become enriched in the liquid phase while the lighter isotopes (16O and 1H) tend toward the vapor phase.

Phase transitions occur when the thermodynamic free energy of a system is non-analytic for some choice of thermodynamic variables (cf. phases). This condition generally stems from the interactions of a large number of particles in a system, and does not appear in systems that are too small. It is important to note that phase transitions can occur and are defined for non-thermodynamic systems, where temperature is not a parameter. Examples include: quantum phase transitions, dynamic phase transitions, and topological (structural) phase transitions. In these types of systems other parameters take the place of temperature. For instance, connection probability replaces temperature for percolating networks.

At the phase transition point (for instance, boiling point) the two phases of a substance, liquid and vapor, have identical free energies and therefore are equally likely to exist. Below the boiling point, the liquid is the more stable state of the two, whereas above the gaseous form is preferred.

It is sometimes possible to change the state of a system diabatically (as opposed to adiabatically) in such a way that it can be brought past a phase transition point without undergoing a phase transition. The resulting state is metastable, i.e., less stable than the phase to which the transition would have occurred, but not unstable either. This occurs in superheating, supercooling, and supersaturation, for example.

Classifications

Ehrenfest Classification

Paul Ehrenfest classified phase transitions based on the behavior of the thermodynamic free energy as a function of other thermodynamic variables. Under this scheme, phase transitions were labeled by the lowest derivative of the free energy that is discontinuous at the transition. *First-order phase transitions* exhibit a discontinuity in the first derivative of the free energy with respect to some thermodynamic variable. The various solid/liquid/gas transitions are classified as first-order transitions because they involve a discontinuous change in density, which is the (inverse of the) first derivative of the free energy with respect to pressure. *Second-order phase transitions* are continuous in the first derivative (the order parameter, which is the first derivative of the free energy with respect to the external field, is continuous across the transition) but exhibit discontinuity in a second derivative of the free energy. These include the ferromagnetic phase transition in materials such as iron, where the magnetization, which is the first derivative of the free energy with respect to the applied magnetic field strength, increases continuously from zero as the temperature is lowered below the Curie temperature. The magnetic susceptibility, the second derivative of the free energy with the field, changes discontinuously. Under the Ehrenfest classification scheme, there could in principle be third, fourth, and higher-order phase transitions.

Though useful, Ehrenfest's classification has been found to be an incomplete method of classifying phase transitions, for it does not take into account the case where a derivative of free energy diverges (which is only possible in the thermodynamic limit).

For instance, in the ferromagnetic transition, the heat capacity diverges to infinity.The same phenomena is also seen in superconducting phase transition.

Modern Classifications

In the modern classification scheme, phase transitions are divided into two broad categories, named similarly to the Ehrenfest classes:

First-order phase transitions are those that involve a latent heat. During such a transition, a system either absorbs or releases a fixed (and typically large) amount of energy per volume. During this process, the temperature of the system will stay constant as heat is added: the system is in a "mixed-phase regime" in which some parts of the system have completed the transition and others have not. Familiar examples are the melting of ice or the boiling of water (the water does not instantly turn into vapor, but forms a turbulent mixture of liquid water and vapor bubbles). Imry and Wortis showed that quenched disorder can broaden a first-order transition in that the transformation is completed over a finite range of temperatures, but phenomena like supercooling and superheating survive and hysteresis is observed on thermal cycling.

Second-order phase transitions are also called *continuous phase transitions*. They are characterized by a divergent susceptibility, an infinite correlation length, and a power-law decay of correlations near criticality. Examples of second-order phase transitions are the ferromagnetic transition, superconducting transition (for a Type-I superconductor the phase transition is second-order at zero external field and for a Type-II superconductor the phase transition is second-order for both normal state-mixed state and mixed state-superconducting state transitions) and the superfluid transition. In contrast to viscosity, thermal expansion and heat capacity of amorphous materials show a relatively sudden change at the glass transition temperature which enables accurate detection using differential scanning calorimetry measurements. Lev Landau gave a phenomenological theory of second-order phase transitions.

Apart from isolated, simple phase transitions, there exist transition lines as well as multicritical points, when varying external parameters like the magnetic field or composition.

Several transitions are known as the *infinite-order phase transitions*. They are continuous but break no symmetries. The most famous example is the Kosterlitz–Thouless transition in the two-dimensional XY model. Many quantum phase transitions, e.g., in two-dimensional electron gases, belong to this class.

The liquid–glass transition is observed in many polymers and other liquids that can be supercooled far below the melting point of the crystalline phase. This is atypical in several respects. It is not a transition between thermodynamic ground states: it is widely believed that the true ground state is always crystalline. Glass is a *quenched disorder* state, and its entropy, density, and so on, depend on the thermal history. Therefore, the glass transition is primarily a dynamic phenomenon: on cooling a liquid, internal degrees of freedom successively fall out of equilibrium. Some theoretical methods predict an underlying phase transition in the hypothetical limit of infinitely long relaxation times. No direct experimental evidence supports the existence of these transitions.

Characteristic Properties

Phase Coexistence

A disorder-broadened first-order transition occurs over a finite range of temperatures where the fraction of the low-temperature equilibrium phase grows from zero to one (100%) as the temperature is lowered. This continuous variation of the coexisting fractions with temperature raised interesting possibilities. On cooling, some liquids vitrify into a glass rather than transform to the equilibrium crystal phase. This happens if the cooling rate is faster than a critical cooling rate, and is attributed to the molecular motions becoming so slow that the molecules cannot rearrange into the crystal positions. This slowing down happens below a glass-formation temperature Tg, which may depend on the applied pressure., If the first-order freezing transition occurs over a range of temperatures, and Tg falls within this range, then there is an interesting possibility that the transition is arrested when it is partial and incomplete. Extending these ideas to first-order magnetic transitions being arrested at low temperatures, resulted in the observation of incomplete magnetic transitions, with two magnetic phases coexisting, down to the lowest temperature. First reported in the case of a ferromagnetic to anti-ferromagnetic transition, such persistent phase coexistence has now been reported across a variety of first-order magnetic transitions. These include colossal-magnetoresistance manganite materials, magnetocaloric materials, magnetic shape memory materials, and other materials. The interesting feature of these observations of Tg falling within the temperature range over which the transition occurs is that the first-order magnetic transition is influenced by magnetic field, just like the structural transition is influenced by pressure. The relative ease with which magnetic field can be controlled, in contrast to pressure, raises the possibility that one can study the interplay between Tg and Tc in an exhaustive way. Phase coexistence across first-order magnetic transitions will then enable the resolution of outstanding issues in understanding glasses.

Critical Points

In any system containing liquid and gaseous phases, there exists a special combination of pressure and temperature, known as the critical point, at which the transition between liquid and gas becomes a second-order transition. Near the critical point, the fluid is sufficiently hot and compressed that the distinction between the liquid and gaseous phases is almost non-existent. This is associated with the phenomenon of critical opalescence, a milky appearance of the liquid due to density fluctuations at all possible wavelengths (including those of visible light).

Symmetry

Phase transitions often involve a symmetry breaking process. For instance, the cooling of a fluid into a crystalline solid breaks continuous translation symmetry: each point in the fluid has the same properties, but each point in a crystal does not have the same properties (unless the points are chosen from the lattice points of the crystal lattice). Typically, the high-temperature phase contains more symmetries than the low-temperature phase due to spontaneous symmetry breaking, with the exception of certain accidental symmetries (e.g. the formation of heavy virtual particles, which only occurs at low temperatures).

Order Parameters

An order parameter is a measure of the degree of order across the boundaries in a phase transition system; it normally ranges between zero in one phase (usually above the critical point) and nonzero in the other. At the critical point, the order parameter susceptibility will usually diverge.

An example of an order parameter is the net magnetization in a ferromagnetic system undergoing a phase transition. For liquid/gas transitions, the order parameter is the difference of the densities.

From a theoretical perspective, order parameters arise from symmetry breaking. When this happens, one needs to introduce one or more extra variables to describe the state of the system. For example, in the ferromagnetic phase, one must provide the net magnetization, whose direction was spontaneously chosen when the system cooled below the Curie point. However, note that order parameters can also be defined for non-symmetry-breaking transitions. Some phase transitions, such as superconducting and ferromagnetic, can have order parameters for more than one degree of freedom. In such phases, the order parameter may take the form of a complex number, a vector, or even a tensor, the magnitude of which goes to zero at the phase transition.

There also exist dual descriptions of phase transitions in terms of disorder parameters. These indicate the presence of line-like excitations such as vortex- or defect lines.

Relevance in Cosmology

Symmetry-breaking phase transitions play an important role in cosmology. It has been speculated by Lee Smolin and Benjamin and Jeremy Bernstein that, in the hot early universe, the vacuum (i.e. the various quantum fields that fill space) possessed a large number of symmetries. As the universe expanded and cooled, the vacuum underwent a series of symmetry-breaking phase transitions. For example, the electroweak transition broke the SU(2)×U(1) symmetry of the electroweak field into the U(1) symmetry of the present-day electromagnetic field. This transition is important to understanding the asymmetry between the amount of matter and antimatter in the present-day universe.

Progressive phase transitions in an expanding universe are implicated in the development of order in the universe, as is illustrated by the work of Eric Chaisson and David Layzer.

Critical Exponents and Universality Classes

Continuous phase transitions are easier to study than first-order transitions due to the absence of latent heat, and they have been discovered to have many interesting properties. The phenomena associated with continuous phase transitions are called critical phenomena, due to their association with critical points.

It turns out that continuous phase transitions can be characterized by parameters known as critical exponents. The most important one is perhaps the exponent describing the divergence of the thermal correlation length by approaching the transition. For instance, let us examine the behavior of the heat capacity near such a transition. We vary the temperature T of the system while keeping all the other thermodynamic variables fixed, and find that the transition occurs at some critical temperature T_c. When T is near T_c, the heat capacity C typically has a power law behavior,

$$C \propto |T_c - T|^{-\alpha}.$$

The heat capacity of amorphous materials has such a behaviour near the glass transition temperature where the universal critical exponent α = 0.59 A similar behavior, but with the exponent ν instead of α, applies for the correlation length.

The exponent ν is positive. This is different with α. Its actual value depends on the type of phase transition we are considering.

It is widely believed that the critical exponents are the same above and below the critical temperature. It has now been shown that this is not necessarily true: When a continuous symmetry is explicitly broken down to a discrete symmetry by irrelevant (in the renormalization group sense) anisotropies, then some exponents (such as γ, the exponent of the susceptibility) are not identical.

For −1 < α < 0, the heat capacity has a "kink" at the transition temperature. This is the behavior of liquid helium at the lambda transition from a normal state to the superfluid state, for which experiments have found α = -0.013±0.003. At least one experiment was performed in the zero-gravity conditions of an orbiting satellite to minimize pressure differences in the sample. This experimental value of α agrees with theoretical predictions based on variational perturbation theory.

For $0 < \alpha < 1$, the heat capacity diverges at the transition temperature (though, since $\alpha < 1$, the enthalpy stays finite). An example of such behavior is the 3D ferromagnetic phase transition. In the three-dimensional Ising model for uniaxial magnets, detailed theoretical studies have yielded the exponent $\alpha \sim +0.110$.

Some model systems do not obey a power-law behavior. For example, mean field theory predicts a finite discontinuity of the heat capacity at the transition temperature, and the two-dimensional Ising model has a logarithmic divergence. However, these systems are limiting cases and an exception to the rule. Real phase transitions exhibit power-law behavior.

Several other critical exponents, β, γ, δ, ν, and η, are defined, examining the power law behavior of a measurable physical quantity near the phase transition. Exponents are related by scaling relations, such as

$$\beta = \gamma / (\delta - 1), \qquad \nu = \gamma / (2 - \eta).$$

It can be shown that there are only two independent exponents, e.g. ν and η.

It is a remarkable fact that phase transitions arising in different systems often possess the same set of critical exponents. This phenomenon is known as *universality*. For example, the critical exponents at the liquid–gas critical point have been found to be independent of the chemical composition of the fluid.

More impressively, but understandably from above, they are an exact match for the critical exponents of the ferromagnetic phase transition in uniaxial magnets. Such systems are said to be in the same universality class. Universality is a prediction of the renormalization group theory of phase transitions, which states that the thermodynamic properties of a system near a phase transition depend only on a small number of features, such as dimensionality and symmetry, and are insensitive to the underlying microscopic properties of the system. Again, the divergence of the correlation length is the essential point.

Critical Slowing Down and other Phenomena

There are also other critical phenomena; e.g., besides *static functions* there is also *critical dynamics*. As a consequence, at a phase transition one may observe critical slowing down or *speeding up*. The large *static universality classes* of a continuous phase transition split into smaller *dynamic universality* classes. In addition to the critical exponents, there are also universal relations for certain static or dynamic functions of the magnetic fields and temperature differences from the critical value.

Percolation Theory

Another phenomenon which shows phase transitions and critical exponents is percolation. The simplest example is perhaps percolation in a two dimensional square lattice.

Sites are randomly occupied with probability p. For small values of p the occupied sites form only small clusters. At a certain threshold p_c a giant cluster is formed and we have a second-order phase transition. The behavior of P_∞ near p_c is, $P_\infty \sim (p-p_c)^\beta$, where β is a critical exponent.

Phase Transitions in Biological Systems

Phase transitions play many important roles in biological systems. Examples include the lipid bilayer formation, the coil-globule transition in the process of protein folding and DNA melting, liquid crystal-like transitions in the process of DNA condensation, and cooperative ligand binding to DNA and proteins with the character of phase transition.

In *biological membranes*, gel to liquid crystalline phase transitions play a critical role in physiological functioning of biomembranes. In gel phase, due to low fluidity of membrane lipid fatty-acyl chains, membrane proteins have restricted movement and thus are restrained in exercise of their physiological role. Plants depend critically on photosynthesis by chloroplast thylakoid membranes which are exposed cold environmental temperatures. Thylakoid membranes retain innate fluidity even at relatively low temperatures because of high degree of fatty-acyl disorder allowed by their high content of linolenic acid, 18-carbon chain with 3-double bonds. Gel-to-liquid crystalline phase transition temperature of biological membranes can be determined by many techniques including calorimetry, flouorescence, spin label electron paramagnetic resonance and NMR by recording measurements of the concerned parameter by at series of sample temperatures. A simple method for its determination from 13-C NMR line intensities has also been proposed.

It has been proposed that some biological systems might lie near critical points. Examples include neural networks in the salamander retina, bird flocks gene expression networks in Drosophila, and protein folding. However, it is not clear whether or not alternative reasons could explain some of the phenomena supporting arguments for criticality. It has also been suggested that biological organisms share two key properties of phase transitions: the change of macroscopic behavior and the coherence of a system at a critical point.

Homogeneous Nucleation

- In homogeneous nucleation the probability of nucleation is same throughout the volume of the parent phase.
- The simplest example of nucleation is solidification of a metal.
- Above melting point T_m, liquid free energy, $G_l < G_s$ (Solid free energy) and free energy change for solidification $\Delta G > 0$.

Below T_m, $\Delta G < 0$ and nuclei of the solid phase form.

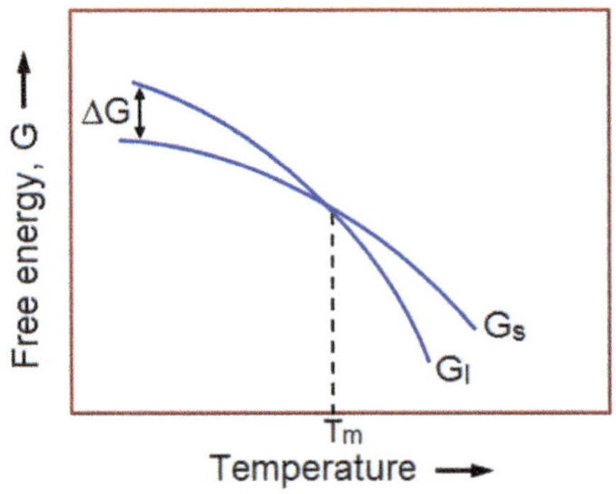

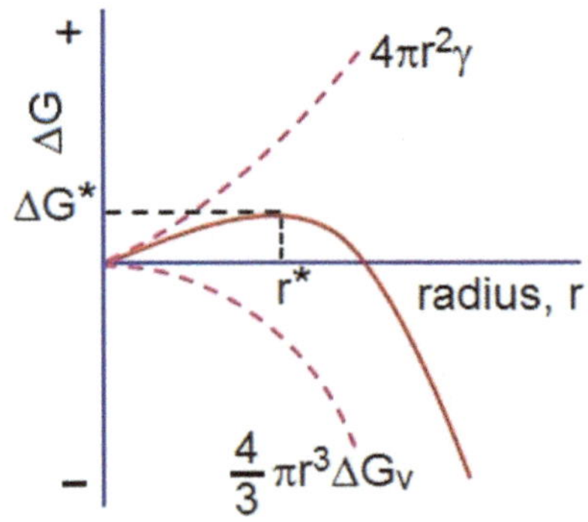

- There are two contributions to free energy change, volume free energy ΔG_v and surface free energy, γ due to creation of a new surface.

- Taking the nucleus as a spherical particle of radius r

$$\Delta G = 4/3\pi r^2 \Delta G_v + 4\pi r^2 \gamma$$

- The tiny particle of the solid that forms first will be stable only when it achieves a critical radius (r*). Below the critical radius it is unstable and is called *embryo*.

- Since this happens at the maximum of the ΔG vs. r curve $d\Delta G / dr = 4\pi r \;\Delta G \; + 8\pi r\gamma = 0$

- This yields $r^* = -\dfrac{2\gamma}{\Delta G_v}$ and $\Delta G^* = \dfrac{16\pi\gamma^3}{3(\Delta G_v)^2}$

- Here, the probability of nucleation is much higher at certain preferred sites such as mold wall, inclusions, grain boundaries, compared to rest of the parent phase.

- Example - Solidification of a liquid on an inclusion surface

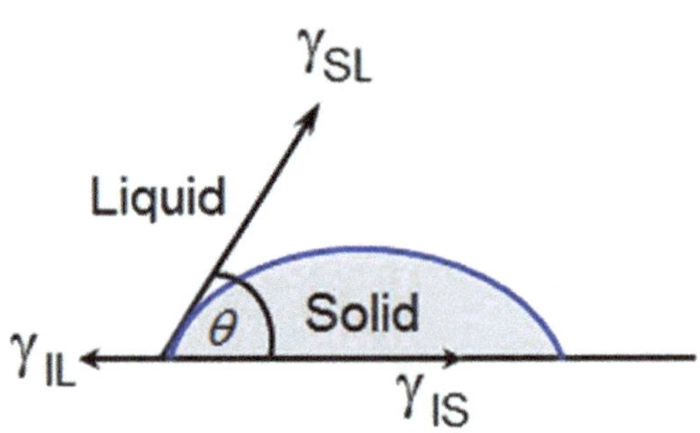

$$\gamma_{IL} = \gamma_{IS} + \gamma_{SL}\cos\theta$$

- With a similar approach it can be shown that $r^* = -\dfrac{2\gamma_{SL}}{\Delta G_v}$ and

$$\Delta G^*_{Het} = \frac{4\pi\gamma^3}{3(\Delta G_v)^2}(2 - 3\cos\theta + \cos^3\theta) = \Delta G^*_{Hom} S(\theta)$$

- The small value of θ ensures that the energy barrier (ΔG) is effectively lowered in heterogeneous nucleation.

Nucleation and Growth Kinetics

- Once the embryo exceeds the critical size r*, the growth of the nucleus starts. Nucleation continues simultaneously.
- Nucleation and growth rates are function of temp. Nucleation rate increases with cooling rate and degree of undercooling ($\Delta T = T_m - T$).
- High nucleation rate and low growth – Finer grain size.
- The over all transformation rate is the product of nucleation and growth rates.

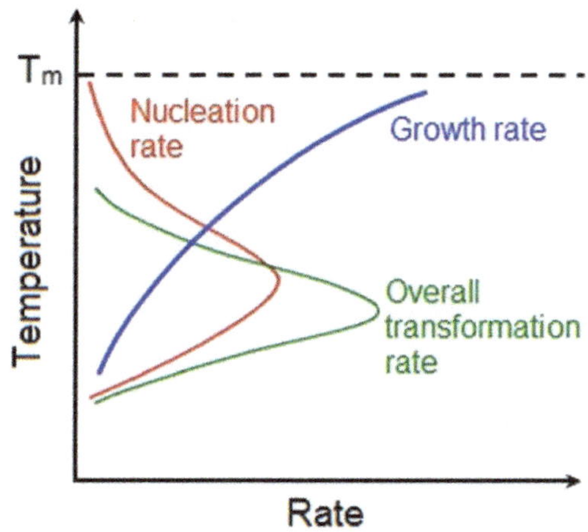

Fe-C Phase Diagram

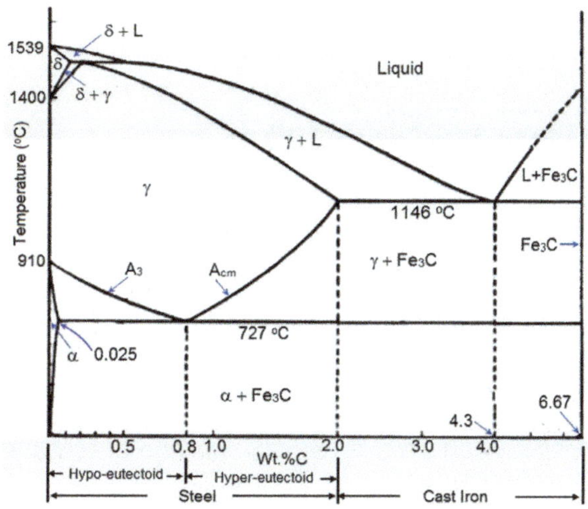

Phases in Fe-C System

- α *-ferrite* – Interstitial solid solution of C in BCC iron. Max solubility of C is 0.025%. Exists from 273°C to 910°C.
- *Austenite* (γ) - Interstitial solid solution of C in FCC iron. Max solubility of C is 2.1%. Exists from 910°C - 1394°C.
- δ *-ferrite* (BCC) exists over the temp range of 1394°C to 1539°C. Max solubility of C is 0.09%.
- Cementite, Fe_3C - is an intermetallic compound. C content in Fe_3C is 6.67%.
- Graphite, the free form of C, also exists in the Fe-C system.
- Bainite (*B*) is another phase which forms in steels at higher cooling rates.
- The hard phase martensite (*M*) forms below the bainitic temperature range at high cooling rates.

Critical Temperatures in Fe-C System

- The eutectoid temperature (727°C) during heating and cooling is A_{c1} and A_{r1} respectively. *A* for arrêt (arrest), *c* for chauffage (heating) and *r* for refroidissement (cooling).
- At normal rates of heating or cooling $A_{c1} > A_{r1}$.
- A_2 at 768 °C is the currie temp above which Fe turns paramagnetic while heating.
- The temperatures corresponding to $(\gamma + \alpha)/\gamma$ and $(\gamma + Fe_3C)/\gamma$ phase boundaries are function of carbon content and are represented as A_3 and A_{cm} respectively.
- The eutectic temperature is 1146 °C.
- The peritectic temperature is at 1495 °C.

Phase Transformation in Fe-C System

- Peritectic reaction at 1495 °C

  L(0.53% C) + δ (0.09% C) → γ (0.17% C)

- Eutectic reaction at 1146°C

  L (4.3% C) → γ (2.1 % C) + Fe_3C (6.67% C). The eutectic mixture of austenite (γ) and cementite (Fe_3C) is called *Ledeburite*. Compositions right and left of 4.3% are called hyper and hypoeutectic steels (Cast iron) respectively.

- Eutectoid reaction at 727°C

  γ (0.8 % C) → α (0.025% C) + Fe_3C (6.67% C). The eutectoid mixture of ferrite (α) and cementite (Fe_3C) is called *Pearlite.*

- Compositions right and left of 0.8% are called hyper and hypoeutectoid steels respectively.

- Compositions up to 2.1% C are steels and beyond this it is considered as cast iron.

Microstructures

- A eutectoid steel (0.8% C) will have 100% pearlite (p) at room temperature (RT). The pearlite formed under equilibrium conditions consists of alternate lamellas of ferrite and Fe_3C.

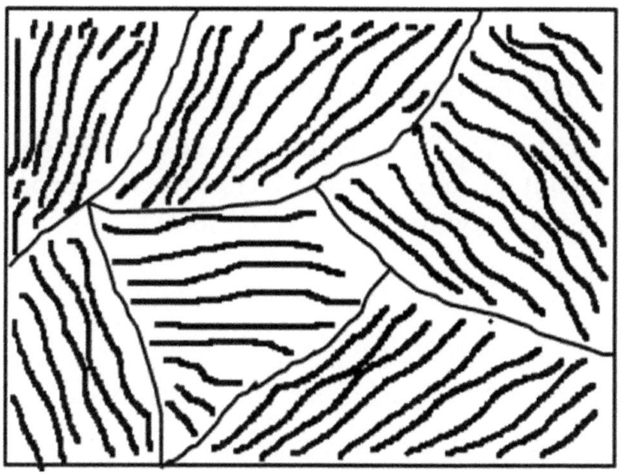

Schematic of Pearlite – White areas are α

- Hypoeutectoid steels – $\alpha + p$; hypereutectoid – $Fe_3C + p$.

- Hypoeutectic cast irons consist of γ + *ledeburite* (Le) below the eutectic temp and $p + Fe_3C + Le$ at RT as the γ transforms to Fe_3C and p at the eutectoid temp. Similarly hyper eutectic cast irons will have a structure of $Fe_3C + Le$.

T-T-T Diagram

- The relation between temperature and time for the formation of a phase is given by T-T-T or temp – time – transformation diagrams also known as isothermal-transformation diagram.

- A typical T-T-T diagram is shown below. The phases formed on isothermal holding at a given temp for a certain period of time are indicated.

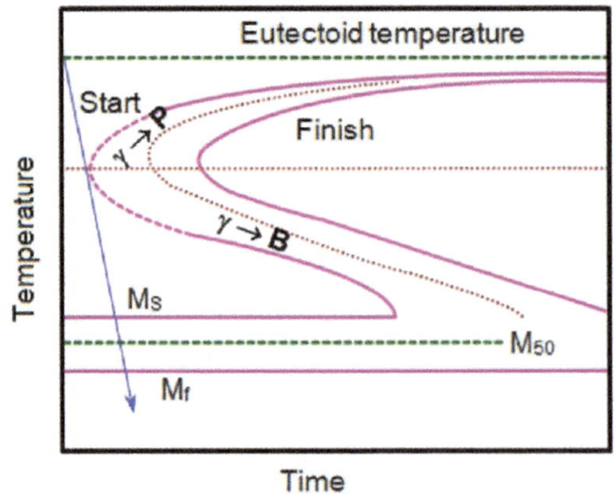

T-T-T diagram

- At normal cooling rates pearlite (*P*) forms, higher cooling rates generates bainite (*B*). The size of pearlite or bainite depends on the transformation temp.
- Martensite (*M*) forms when the steel is cooled below the maternsite start (M_S) temp at much higher cooling rate so that the nose of the T-T-T curve (shown dotted) is avoided (the long blue arrow) .
- Diffusion rates below M_S is so low that $\gamma \rightarrow M$ transformation is a diffusionless process (the C content remains same). However, the crystal structure changes from FCC (γ) to body centered tetragonal (BCT).

C-C-T Diagram

- In actual practice a steel is generally cooled continuously. Continuous-cooling-transformation (C-C-T) diagrams depict this situation.
- The C-C-T curve (Blue) is shifted to the right of the T-T-T (dashed) curve as continuous cooling transformation occurs at lower temperature and longer time compared isothermal holding.

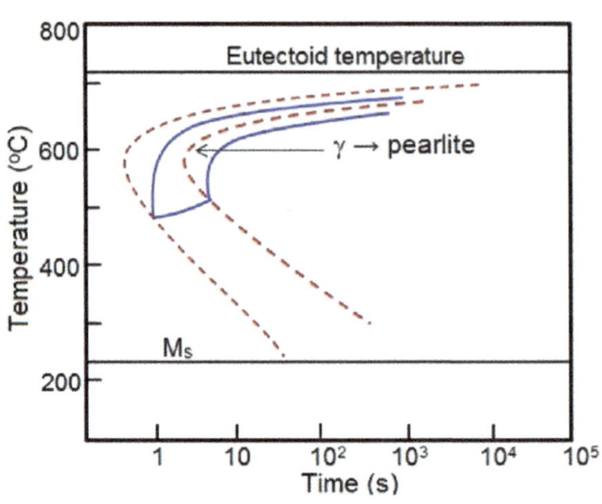

- Bainite generally does not form in steels during continuous cooling and hence the C-C-T curve ceases just below the nose.
- The microstructure (fine or coarse) depends on the cooling rate. Higher the cooling rate finer the microstructure is.
- Finer size pearlite is called *sorbite* and very fine size pearlite is called *troostite*.
- The critical cooling rate is the one at which the cooling curve just touches the nose of the C-C-T curve.
- A cooling rate higher than the critical rate is needed to form martensite.

Diffusionless Transformation

A diffusionless transformation is a phase change that occurs without the long-range diffusion of atoms but rather by some form of cooperative, homogeneous movement of many atoms that results in a change in crystal structure. These movements are small, usually less than the interatomic distances, and the atoms maintain their relative relationships. The ordered movement of large numbers of atoms lead some to refer to these as *military* transformations in contrast to *civilian* diffusion-based phase changes.

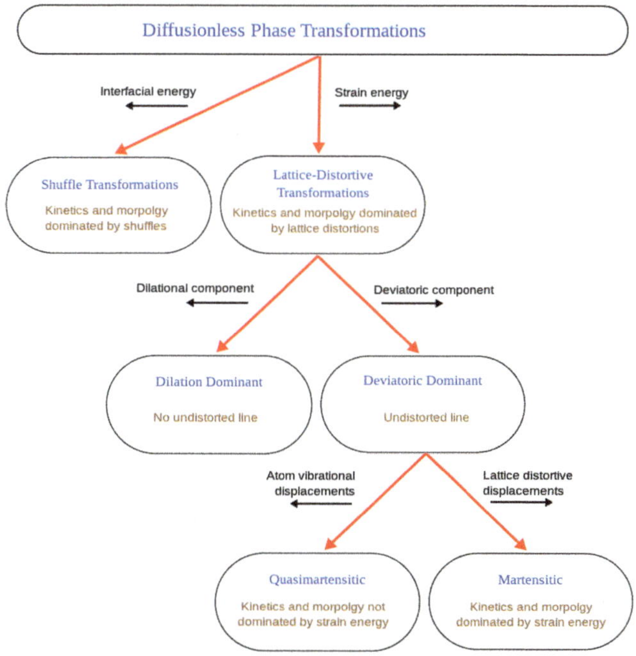

Diffusionless transformations classifications

The most commonly encountered transformation of this type is the martensitic transformation which, while being the least studied, is only one subset of non-diffusional transformations. The martensitic transformation in steel represents the most economically significant example of this category of phase transformations but an increasing number of alternatives, such as shape memory alloys, are becoming more important as well.

Classification and Definitions

When a structural change occurs by the coordinated movement of atoms (or groups of atoms) relative to their neighbors then the change is termed *displacive* transformation. This covers a broad range of transformations and so further classifications have been developed [Cohen 1979].

The first distinction can be drawn between transformations dominated by *lattice-distortive strains* and those where *shuffles* are of greater importance.

Homogeneous lattice-distortive strains, also known as Bain strains, are strains that transform one Bravais lattice into a different one. This can be represented by a strain matrix S which transforms one vector, y, into a new vector, x:

$$y = Sx$$

This is homogeneous as straight lines are transformed to new straight lines. Examples of such transformations include a cubic lattice increasing in size on all three axes (dilation) or shearing into a monoclinic structure.

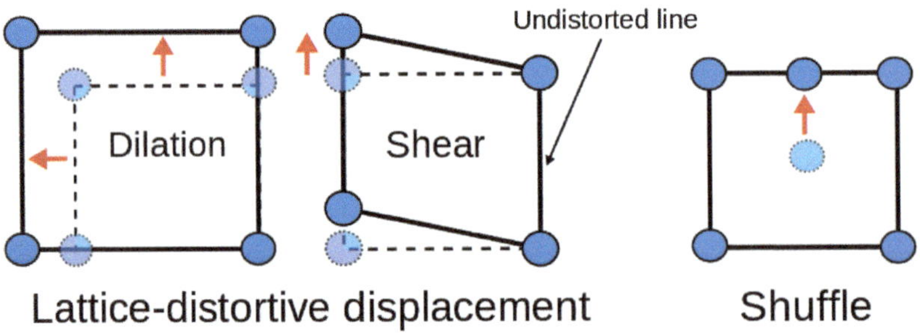

Shuffles, as the name suggests, involve the small movement of atoms within the unit cell. As a result pure shuffles do not normally result in a shape change of the unit cell - only its symmetry and structure.

Phase transformations normally result in the creation of an interface between the transformed and parent material. The energy required to generate this new interface will depend on its nature - essentially how well the two structures fit together. An additional energy term occurs if the transformation includes a shape change since, if new phase is constrained by surrounding material, this may give rise to elastic or plastic deformation and hence a strain energy term. The ratio of these interfacial and strain energy terms has a notable effect on the kinetics of the transformation and the morphology of the new phase. Thus, shuffle transformations, where distortions are small, are dominated by interfacial energies and can be usefully separated from lattice-distortive transformations where the strain energy tends to have a greater effect.

A subclassification of lattice-distortive displacements can be made by considering the dilational and shear components of the distortion. In transformations dominated by the shear component it is possible to find a line in the new phase that is undistorted from the parent phase while all lines are distorted when the dilation is predominant. Shear dominated transformations can be further classified according to the magnitude of the strain energies involved compared to the innate vibrations of the atoms in the lattice and hence whether the strain energies have a notable influence on the kinetics of the transformation and the morphology of the resulting phase. If the strain energy is a significant factor then the transformations are dubbed *martensitic* and if it is not the transformation is referred to as *quasi-martensitic*.

Martensitic Transformation

The difference between austenite and martensite is, in some ways, quite small: while the unit cell of austenite is, on average, a perfect cube, the transformation to martensite distorts this cube by interstitial carbon atoms that do not have time to diffuse out during displacive transformation. The unit cell becomes slightly longer in one dimension and shorter in the other two. The mathematical description of the two structures is quite different, for reasons of symmetry, but the chemical bonding remains very similar. Unlike cementite, which has bonding reminiscent of ceramic materials, the hardness of martensite is difficult to explain in chemical terms.

The explanation hinges on the crystal's subtle change in dimension. Even a microscopic crystallite is millions of unit cells long. Since all of these units face the same direction, distortions of even a fraction of a percent become magnified into a major mismatch between neighboring materials. The mismatch is sorted out by the creation of myriad crystal defects, in a process reminiscent of work hardening. As in work-hardened steel, these defects prevent atoms from sliding past one another in an organized fashion, causing the material to become harder.

Shape memory alloys also have surprising mechanical properties, that were eventually explained by an analogy to martensite. Unlike the iron-carbon system, alloys in the nickel-titanium system can be chosen that make the "martensitic" phase thermodynamically stable.

Pseudomartensitic Transformation

In addition to displacive transformation and diffusive transformation, a new phase transformation that involves displasive sublattice transition and atomic diffusion was discovered using a high-pressure x-ray diffraction system. The new transformation mechanism has been christened a pseudomartensitic transformation.

References

- Banerjee, A; Pramanik, A K; Kumar, Kranti; Chaddah, P (2006). "Coexisting tunable fractions of glassy and equilibrium long-range-order phases in manganites". Journal of Physics: Condensed Matter. 18 (49): L605. Bibcode:2006JPCM...18L.605B. arXiv:cond-mat/0611152 . doi:10.1088/0953-8984/18/49/L02
- Modell, Michael; Robert C. Reid (1974). Thermodynamics and Its Applications. Englewood Cliffs, NJ: Prentice-Hall. ISBN 0-13-914861-2
- Jaeger, Gregg (1 May 1998). "The Ehrenfest Classification of Phase Transitions: Introduction and Evolution". Archive for History of Exact Sciences. 53 (1): 51–81. doi:10.1007/s004070050021
- Lakhani, Archana; Banerjee, A; Chaddah, P; Chen, X; Ramanujan, R V (2012). "Magnetic glass in shape memory alloy: Ni45Co5Mn38Sn12". Journal of Physics: Condensed Matter. 24 (38): 386004. Bibcode:2012JPCM...24L6004L. ISSN 0953-8984. arXiv:1206.2024 . doi:10.1088/0953-8984/24/38/386004
- Blundell, Stephen J.; Katherine M. Blundell (2008). Concepts in Thermal Physics. Oxford University Press. ISBN 978-0-19-856770-7
- Leonard, F.; Delamotte, B. (2015). "Critical exponents can be different on the two sides of a transition". Phys. Rev. Lett. 115 (20): 200601. Bibcode:2015PhRvL.115t0601L. arXiv:1508.07852 . doi:10.1103/PhysRevLett.115.200601
- Ojovan, Michael I; Lee, William E (2006). "Topologically disordered systems at the glass transition" (PDF). Journal of Physics: Condensed Matter. 18 (50): 11507–11520. Bibcode:2006JPCM...1811507O. doi:10.1088/0953-8984/18/50/007
- Ivancevic, Vladimir G.; Ivancevic, Tijiana, T. (2008). Complex Nonlinearity. Berlin: Springer. pp. 176–177. ISBN 978-3-540-79357-1. Retrieved 12 October 2014
- Krotov, D; Dubuis, J O; Gregor, T; Bialek, W (2014). "Morphogenesis at criticality". PNAS. 111: 3683–3688. doi:10.1073/pnas.1324186111
- YashRoy, R.C. (1987). "13-C NMR studies of lipid fatty acyl chains of chloroplast membranes". Indian Journal of Biochemistry and Biophysics. 24 (6): 177–178
- A. D. McNaught and A. Wilkinson (ed.). "Compendium of Chemical Terminology". IUPAC. ISBN 0-86542-684-8. Retrieved 2007-10-23
- Mora, Thierry; Bialek, William (2011). "Are biological systems poised at criticality?". Journal of Statistical Physics. 144 (2): 268–302. Bibcode:2011JSP...144..268M. arXiv:1012.2242 . doi:10.1007/s10955-011-0229-4

4

Properties of Materials

Hardness is the most common mechanical property of a material used to classify and distinguish materials. It is the resistance of a material against any pressure that may lead to any deformation. This section elucidates the crucial theories and principles of the mechanical properties of materials and strengthening mechanisms.

Hardness

Hardness is a measure of how resistant solid matter is to various kinds of permanent shape change when a compressive force is applied. Some materials (e.g. metals) are harder than others (e.g. plastics). Macroscopic hardness is generally characterized by strong intermolecular bonds, but the behavior of solid materials under force is complex; therefore, there are different measurements of hardness: *scratch hardness*, *indentation hardness*, and *rebound hardness*.

Hardness is dependent on ductility, elastic stiffness, plasticity, strain, strength, toughness, viscoelasticity, and viscosity.

Common examples of hard matter are ceramics, concrete, certain metals, and superhard materials, which can be contrasted with soft matter.

Measuring Hardness

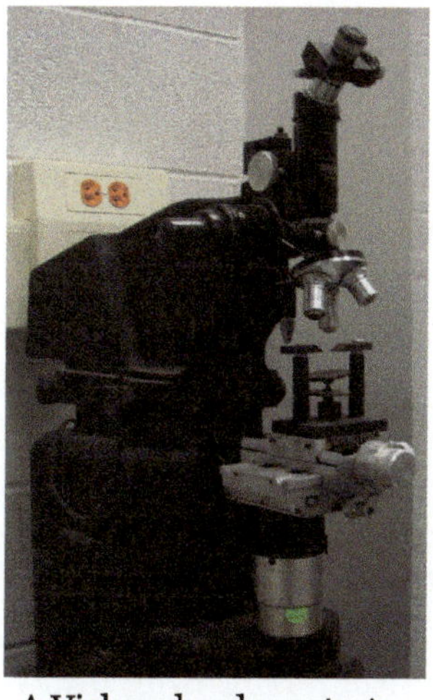

A Vickers hardness tester

There are three main types of hardness measurements: *scratch*, *indentation*, and *rebound*. Within each of these classes of measurement there are individual measurement scales. For practical reasons conversion tables are used to convert between one scale and another.

Scratch Hardness

Scratch hardness is the measure of how resistant a sample is to fracture or permanent plastic deformation due to friction from a sharp object. The principle is that an object made of a harder material will scratch an object made of a softer material. When testing coatings, scratch hardness refers to the force necessary to cut through the film to the substrate. The most common test is Mohs scale, which is used in mineralogy. One tool to make this measurement is the sclerometer.

Another tool used to make these tests is the pocket hardness tester. This tool consists of a scale arm with graduated markings attached to a four-wheeled carriage. A scratch tool with a sharp rim is mounted at a predetermined angle to the testing surface. In order to use it a weight of known mass is added to the scale arm at one of the graduated markings, the tool is then drawn across the test surface. The use of the weight and markings allows a known pressure to be applied without the need for complicated machinery.

Indentation Hardness

Indentation hardness measures the resistance of a sample to material deformation due to a constant compression load from a sharp object; they are primarily used in engineering and metallurgy fields. The tests work on the basic premise of measuring the critical dimensions of an indentation left by a specifically dimensioned and loaded indenter.

Common indentation hardness scales are Rockwell, Vickers, Shore, and Brinell.

Rebound Hardness

Rebound hardness, also known as *dynamic hardness*, measures the height of the "bounce" of a diamond-tipped hammer dropped from a fixed height onto a material. This type of hardness is related to elasticity. The device used to take this measurement is known as a scleroscope.

Two scales that measures rebound hardness are the Leeb rebound hardness test and Bennett hardness scale.

Hardening

There are five hardening processes: Hall-Petch strengthening, work hardening, solid solution strengthening, precipitation hardening, and martensitic transformation.

Physics

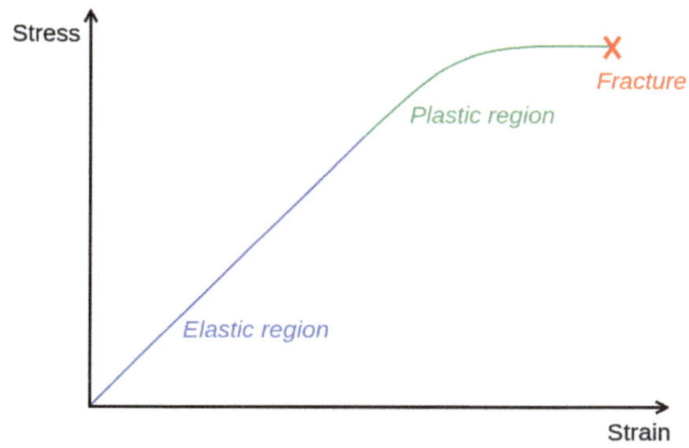

Diagram of a stress-strain curve, showing the relationship between stress (force applied per unit area) and strain or deformation of a ductile metal.

In solid mechanics, solids generally have three responses to force, depending on the amount of force and the type of material:

- They exhibit elasticity—the ability to temporarily change shape, but return to the original shape when the pressure is removed. "Hardness" in the elastic range—a small temporary change in shape for a given force—is known as stiffness in the case of a given object, or a high elastic modulus in the case of a material.
- They exhibit plasticity—the ability to permanently change shape in response to the force, but remain in one piece. The yield strength is the point at which elastic deformation gives way to plastic deformation. Deformation in the plastic range is non-linear, and is described by the stress-strain curve. This response produces the observed properties of scratch and indentation hardness, as described and measured in materials science. Some materials exhibit both elasticity and viscosity when undergoing plastic deformation; this is called viscoelasticity.
- They fracture—split into two or more pieces.

Strength is a measure of the extent of a material's elastic range, or elastic and plastic ranges together. This is quantified as compressive strength, shear strength, tensile strength depending on the direction of the forces involved. Ultimate strength is an engineering measure of the maximum load a part of a specific material and geometry can withstand.

Brittleness, in technical usage, is the tendency of a material to fracture with very little or no detectable plastic deformation beforehand. Thus in technical terms, a material can be both brittle and strong. In everyday usage "brittleness" usually refers to the tendency to fracture under a small amount of force, which exhibits both brittleness and a lack of strength (in the technical sense). For perfectly brittle materials, yield strength and ultimate strength are the same, because they do not experience detectable plastic deformation. The opposite of brittleness is ductility.

The toughness of a material is the maximum amount of energy it can absorb before fracturing, which is different from the amount of force that can be applied. Toughness tends to be small for brittle materials, because elastic and plastic deformations allow materials to absorb large amounts of energy.

Hardness increases with decreasing particle size. This is known as the Hall-Petch relationship. However, below a critical grain-size, hardness decreases with decreasing grain size. This is known as the inverse Hall-Petch effect.

Hardness of a material to deformation is dependent on its microdurability or small-scale shear modulus in any direction, not to any rigidity or stiffness properties such as its bulk modulus or Young's modulus. Stiffness is often confused for hardness. Some materials are stiffer than diamond (e.g. osmium) but are not harder, and are prone to spalling and flaking in squamose or acicular habits.

Mechanisms and Theory

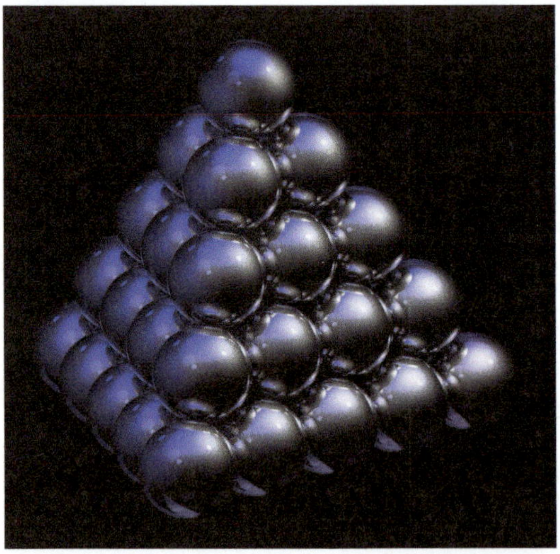

A representation of the crystal lattice showing the planes of atoms.

The key to understanding the mechanism behind hardness is understanding the metallic microstructure, or the structure and arrangement of the atoms at the atomic level. In fact, most important metallic properties critical to the manufacturing of today's goods are determined by the microstructure of a material. At the atomic level, the atoms in a metal are arranged in an orderly three-dimensional array called a crystal lattice. In reality, however, a given specimen of a metal likely never contains a consistent single crystal lattice. A given sample of metal will contain many grains, with each grain having a fairly consistent array pattern. At an even smaller scale, each grain contains irregularities.

There are two types of irregularities at the grain level of the microstructure that are responsible for the hardness of the material. These irregularities are point defects and line defects. A point defect is an irregularity located at a single lattice site inside of the overall three-dimensional lattice of the grain. There are three main point defects. If there is an atom missing from the array, a vacancy defect is formed. If there is a different type of atom at the lattice site that should normally be occupied by a metal atom, a substitutional defect is formed. If there exists an atom in a site where there should normally not be, an interstitial defect is formed. This is possible because space exists between atoms in a crystal lattice. While point defects are irregularities at a single site in the crystal lattice, line defects are irregularities on a plane of atoms. Dislocations are a type of line defect involving the misalignment of these planes. In the case of an edge dislocation, a half plane of atoms is wedged between two planes of atoms. In the case of a screw dislocation two planes of atoms are offset with a helical array running between them.

In glasses, hardness seems to depend linearly on the number of topological constraints acting between the atoms of the network. Hence, the rigidity theory has allowed predicting hardness values with respect to composition.

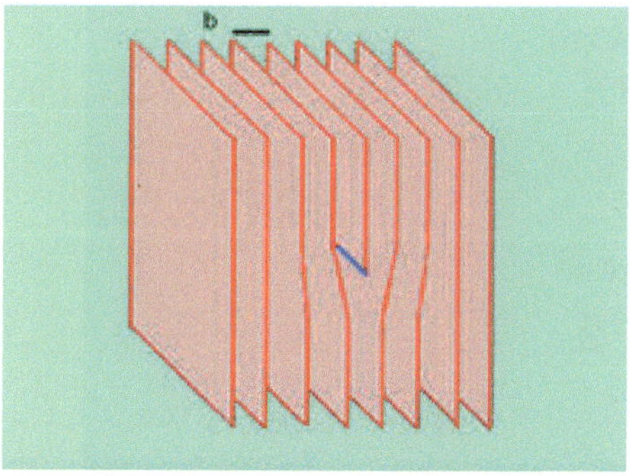

Planes of atoms split by an edge dislocation.

Dislocations provide a mechanism for planes of atoms to slip and thus a method for plastic or permanent deformation. Planes of atoms can flip from one side of the dislocation to the other effectively allowing the dislocation to traverse through the material and the material to deform permanently. The movement allowed by these dislocations causes a decrease in the material's hardness.

The way to inhibit the movement of planes of atoms, and thus make them harder, involves the interaction of dislocations with each other and interstitial atoms. When a dislocation intersects with a second dislocation, it can no longer traverse through the crystal lattice. The intersection of dislocations creates an anchor point and does not allow the planes of atoms to continue to slip over one another A dislocation can also be anchored by the interaction with interstitial atoms. If a dislocation comes in contact with two or more interstitial atoms, the slip of the planes will again be disrupted. The interstitial atoms create anchor points, or pinning points, in the same manner as intersecting dislocations.

By varying the presence of interstitial atoms and the density of dislocations, a particular metal's hardness can be controlled. Although seemingly counter-intuitive, as the density of dislocations increases, there are more intersections created and consequently more anchor points. Similarly, as more interstitial atoms are added, more pinning points that impede the movements of dislocations are formed. As a result, the more anchor points added, the harder the material will become.

Mohs Scale of Hardness

Mohs hardness	Mineral	Chemical formula	Absolute hardness
1	Talc	$Mg_3Si_4O_{10}(OH)_2$	1
2	Gypsum	$CaSO_4 \cdot 2H_2O$	3
3	Calcite	$CaCO_3$	9
4	Fluorite	CaF_2	21
5	Apatite	$Ca_5(PO_4)_3(OH^-,Cl^-,F^-)$	48
6	Orthoclase Feldspar	$KAlSi_3O_8$	72
7	Quartz	SiO_2	100
8	Topaz	$Al_2SiO_4(OH^-,F^-)_2$	200
9	Corundum	Al_2O_3	400
10	Diamond	C	1600

Rockwell Hardness

- In this type of test, depth of indentation at a constant load is taken as the measure of Hardness.
- Aminor load of 10 kg is first applied for good contact between the indenter and the sample surface.
- The major load is then applied and the depth of indentation is recorded on a dial gage in terms of an arbitrary number.
- The dial consists of 100 divisions, each division representing a penetration depth of 0.002 mm.

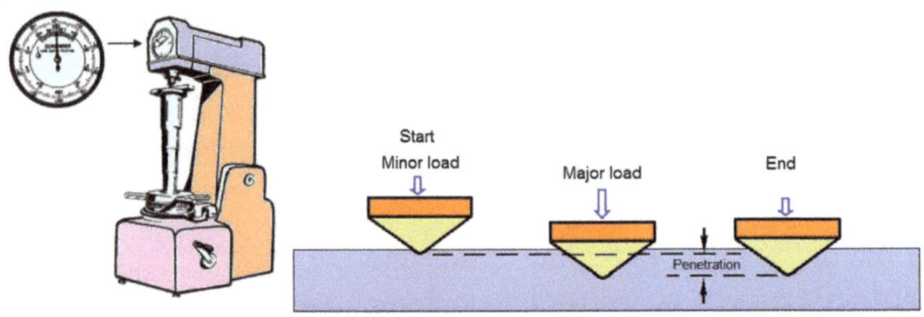

Indenter and Hardness Scale

- Two types of indenters – 120° diamond cone called Brale indenter and 1.6 and 3.2 mm diameter steel balls
- Combination of indenter and major load gives rise to different hardness scales.
- C - Scale – Brale indenter + 150 kg load, designated as R_C. Range is R_C 20 – R_C 70. Used for hard materials like hardened steels.
- B-Scale – Steel ball indenter + 100 kg load, written as R_B. Range is R_B 0 to R_B 100.
- Minor loads in R_C and R_B scales are 10 kg and 3 kg respectively.

Brinell Hardness

- Indentation is done with 10 mm diameter steel ball.
- Aload of 3000 kg (500 kg for softer materials) is applied for 10 – 30 s.
- Dia of the indentation is measured to obtain the hardness (Brinell Hardness No.) from the relationship

$$BHN = \frac{P}{\left(\frac{\pi D}{2}\right)\left(D - \sqrt{D^2 - d^2}\right)} = \frac{P}{\pi D t} \; (kgf / m^2)$$

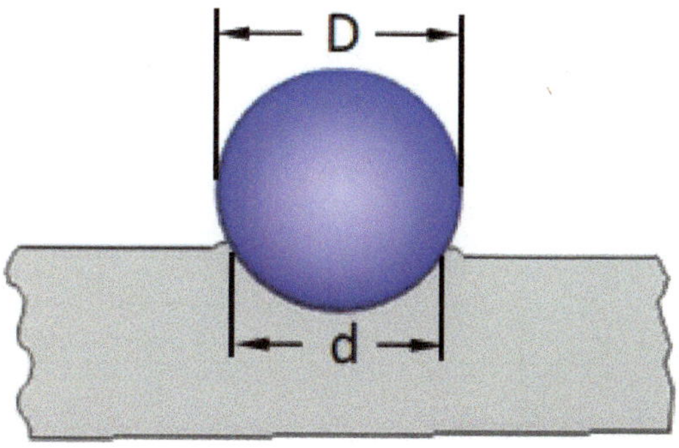

P = Applied load

D = Diameter of ball

d = Dia of indentation

t = Depth of impression

BHN varies with load. P/D^2 value needs to be kept constant according to below euation. $P_1/D_1^2 = P_2/D_2^2 = P_3/D_3^2$

$$BHN = \frac{P}{\left(\frac{\pi}{2}\right) D^2 (1 - \cos\phi)}$$

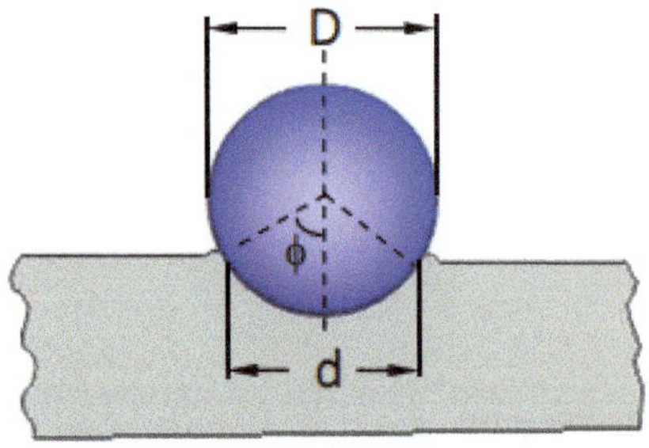

$$d = D \sin\phi$$

Vickers Hardness

- Vickers test uses a square-base diamond pyramid indenter having an angle of 136° between the opposite faces. This angle approximates the ideal d/D ratio (0.375) in Brinell test (Fig. a)
- The hardness, called DPH or VHN (Diamond pyramid hardness no. or Vickers Hardness no.), is obtained by dividing the load (1 – 120 kg) with the surface area of the indentation.
- The surface area is calculated from the diagonals length of the impression.

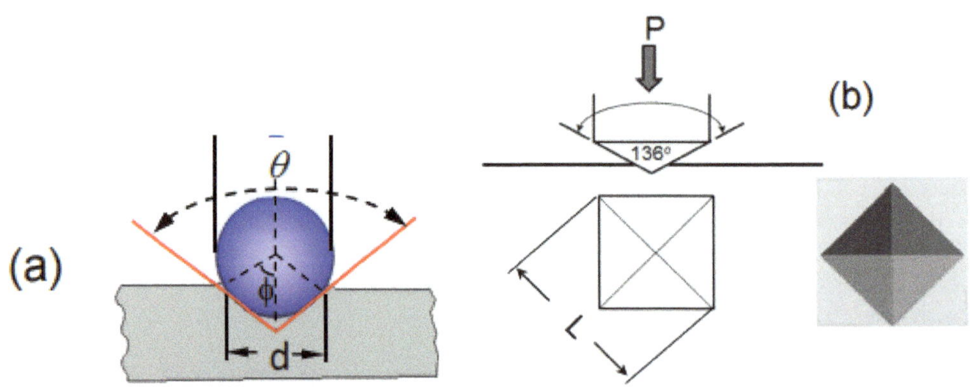

$$DPH(VHN) = \frac{2P\sin(\theta/2)}{L^2} = \frac{1.854P}{L^2}$$

Microhardness

- Sometime hardness determination is needed over a very small area.

- For example, hardness of carburised steel surface, coatings or individual phases or constituents of a material.
- The load applied is much smaller compared to macrohardness.
- The indentation is very small. An optical microscope is used to observe it. Sample preparation is needed.
- Two methods are used for microhardness testing.

Knoop Indentation

- Knoop indenter is a diamond pyramidal indenter. Produces diamond shaped indentation with long and short diagonal lengths in the ratio of 7:1
- The hardness is called Knoop Hardness number (KHN) and is obtained by dividing the load (25 - 300 g) with the projected surface area of the indentation.

$$KHN = \frac{P}{A_p} = \frac{P}{L^2C} \ (kgf / m^2)$$

$$A_p = \text{Projected area of identation}$$

$$L = \text{Longer diagonal length}$$

$$C = \text{Identer specific constant}$$

Vickers Microhardness

- This is same as Vickers hardness except that the applied load is much smaller so as to cover a small area.
- The applied load range is 1 – 100 g.

Ultimate Tensile Strength

Ultimate tensile strength (UTS), often shortened to tensile strength (TS) or ultimate strength, is the capacity of a material or structure to withstand loads tending to elongate, as opposed to compressive strength, which withstands loads tending to reduce size. In other words, tensile strength resists tension (being pulled apart), whereas compressive strength resists compression (being pushed together). Ultimate tensile strength is measured by the maximum stress that a material can withstand while being stretched or pulled before breaking. In the study of strength of materials, tensile strength, compressive strength, and shear strength can be analyzed independently.

Some materials break very sharply, without plastic deformation, in what is called a brittle failure. Others, which are more ductile, including most metals, experience some plastic deformation and possibly necking before fracture.

Two vises apply tension to a specimen by pulling at it, stretching the specimen until it fractures. The maximum stress it withstands before fracturing is its ultimate tensile strength.

The UTS is usually found by performing a tensile test and recording the engineering stress versus strain. The highest point of the stress–strain curve is the UTS. It is an intensive property; therefore its value does not depend on the size of the test specimen. However, it is dependent on other factors, such as the preparation of the specimen, the presence or otherwise of surface defects, and the temperature of the test environment and material.

Tensile strengths are rarely used in the design of ductile members, but they are important in brittle members. They are tabulated for common materials such as alloys, composite materials, ceramics, plastics, and wood.

Tensile strength can be defined for liquids as well as solids under certain conditions. For example, when a tree draws water from its roots to its upper leaves by transpiration, the column of water is pulled upwards from the top by the cohesion of the water in the xylem, and this force is transmitted down the column by its tensile strength. Air pressure, osmotic pressure, and capillary tension also plays a small part in a tree's ability to draw up water, but this alone would only be sufficient to push the column of water to a height of less than ten metres, and trees can grow much higher than that (over 100 m).

Tensile strength is defined as a stress, which is measured as force per unit area. For some non-homogeneous materials (or for assembled components) it can be reported just as a force or as a force per unit width. In the International System of Units (SI), the unit is the pascal (Pa) (or a multiple thereof, often megapascals (MPa), using the SI prefix *mega*); or, equivalently to pascals, newtons per square metre (N/m²). A United States customary unit is pounds per square inch (lb/in² or psi), or kilo-pounds per square inch (ksi, or sometimes kpsi), which is equal to 1000 psi; kilo-pounds per square inch are commonly used in one country (US), when measuring tensile strengths.

Concept

Ductile Materials

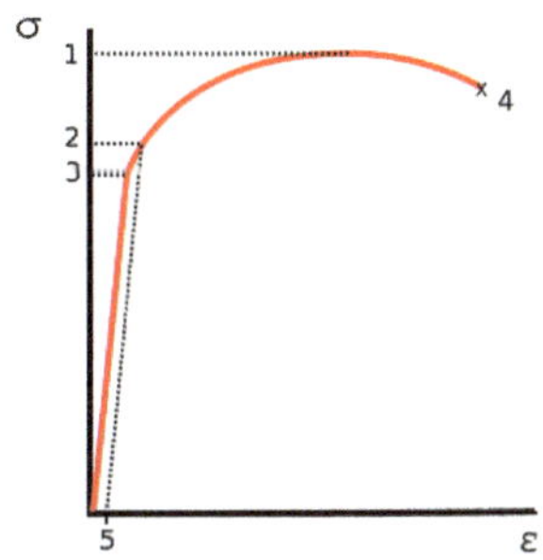

"Engineering" stress–strain (σ–ε) curve typical of aluminum
1. Ultimate strength 2. Yield strength 3. Proportional limit stress
4. Fracture 5. Offset strain (typically 0.2%)

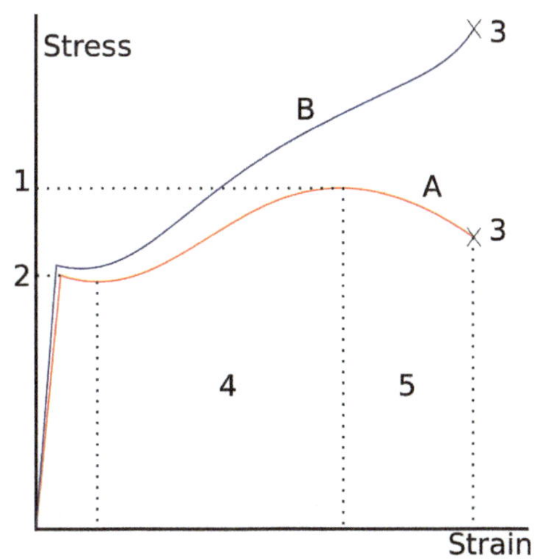

"Engineering" (red) and "true" (blue) stress–strain curve typical of structural steel.
1: Ultimate strength 2: Yield strength (yield point) 3: Rupture
4: Strain hardening region 5: Necking region A: Apparent stress (F/A_0) B: Actual stress (F/A)

Many materials can display linear elastic behavior, defined by a linear stress–strain relationship, as shown in the left figure up to point 3. The elastic behavior of materials often extends into a non-linear region, represented in the figure by point 2 (the "yield point"), up to which deformations are completely recoverable upon removal of the load; that is, a specimen loaded elastically in tension will elongate, but will return to its original shape and size when unloaded. Beyond this elastic region, for ductile materials, such as steel, deformations are plastic. A plastically deformed specimen does not completely return to its original size and shape when unloaded.

For many applications, plastic deformation is unacceptable, and is used as the design limitation.

After the yield point, ductile metals undergo a period of strain hardening, in which the stress increases again with increasing strain, and they begin to neck, as the cross-sectional area of the specimen decreases due to plastic flow. In a sufficiently ductile material, when necking becomes substantial, it causes a reversal of the engineering stress–strain curve (curve A, right figure); this is because the *engineering stress* is calculated assuming the original cross-sectional area before necking. The reversal point is the maximum stress on the engineering stress–strain curve, and the engineering stress coordinate of this point is the ultimate tensile strength, given by point 1.

The UTS is not used in the design of ductile static members because design practices dictate the use of the yield stress. It is, however, used for quality control, because of the ease of testing. It is also used to roughly determine material types for unknown samples.

The UTS is a common engineering parameter to design members made of brittle material because such materials have no yield point.

Testing

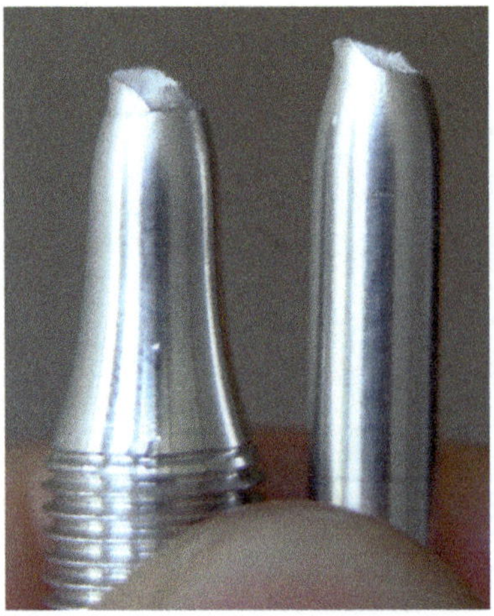

Round bar specimen after tensile stress testing

Typically, the testing involves taking a small sample with a fixed cross-sectional area, and then pulling it with a tensometer at a constant strain (change in gauge length divided by initial gauge length) rate until the sample breaks.

When testing some metals, indentation hardness correlates linearly with tensile strength. This important relation permits economically important nondestructive testing of bulk metal deliveries with lightweight, even portable equipment, such as handheld Rockwell hardness testers. This practical correlation helps quality assurance in metalworking industries to extend well beyond the laboratory and universal testing machines.

While most metal forms, such as sheet, bar, tube, and wire, can exhibit the test UTS, fibers, such as carbon fibers, being only 2/10,000th of an inch in diameter, must be made into composites to create useful real-world forms. As the datasheet on T1000G below indicates, while the UTS of the fiber is very high at 6,370 MPa, the UTS of a derived composite is 3,040 MPa - less than half the strength of the fiber.

Typical Tensile Strengths

Typical tensile strengths of some materials			
Material	**Yield strength (MPa)**	**Ultimate tensile strength (MPa)**	**Density (g/cm³)**
Steel, structural ASTM A36 steel	250	400-550	7.8
Steel, 1090 mild	247	841	7.58
Chromium-vanadium steel AISI 6150	620	940	7.8
Human skin	15	20	2
Steel, 2800 Maraging steel	2617	2693	8.00
Steel, AerMet 340	2160	2430	7.86
Steel, Sandvik Sanicro 36Mo logging cable precision wire	1758	2070	8.00
Steel, AISI 4130, water quenched 855 °C (1570 °F), 480 °C (900 °F) temper	951	1110	7.85
Steel, API 5L X65	448	531	7.8
Steel, high strength alloy ASTM A514	690	760	7.8
Acrylic, clear cast sheet (PMMA)	72	87	1.16
High-density polyethylene (HDPE)	26-33	37	0.85
Polypropylene	12-43	19.7-80	0.91
Steel, stainless AISI 302 - cold-rolled	520	860	8.19
Cast iron 4.5% C, ASTM A-48	130	200	
"Liquidmetal" alloy	1723	550-1600	6.1
Beryllium 99.9% Be	345	448	1.84
Aluminium alloy 2014-T6	414	483	2.8
Polyester resin (unreinforced)	55	55	
Polyester and chopped strand mat laminate 30% E-glass	100	100	
S-Glass epoxy composite	2358	2358	
Aluminium alloy 6061-T6	241	300	2.7
Copper 99.9% Cu	70	220	8.92
Cupronickel 10% Ni, 1.6% Fe, 1% Mn, balance Cu	130	350	8.94

Brass	200 +	550	8.73
Tungsten	941	1510	19.25
Glass		33	2.53
E-Glass	N/A	1500 for laminates, 3450 for fibers alone	2.57
S-Glass	N/A	4710	2.48
Basalt fiber	N/A	4840	2.7
Marble	N/A	15	2.6
Concrete	N/A	2-5	2.7
Carbon fiber	N/A	1600 for laminates, 4137 for fibers alone	1.75
Carbon fiber (Toray T1000G) (the strongest man-made fibres)		6370 fibre alone	1.80
Human hair		10	
Bamboo		350-500	0.4
Spider silk		1000	1.3
Spider silk, Darwin's bark spider	1652		
Silkworm silk	500		1.3
Aramid (Kevlar or Twaron)	3620	3757	1.44
UHMWPE	24	52	0.97
UHMWPE fibers (Dyneema or Spectra)		2300-3500	0.97
Vectran		2850-3340	
Polybenzoxazole (Zylon)	2700	5800	1.56
Wood, pine (parallel to grain)		40	
Bone (limb)	104-121	130	1.6
Nylon, molded, type 6/6	45	75	1.15
Nylon fiber, drawn		900	1.13
Epoxy adhesive	-	12 - 30	-
Rubber	-	16	
Boron	N/A	3100	2.46
Silicon, monocrystalline (m-Si)	N/A	7000	2.33
Ultra-pure silica glass fiber-optic strands		4100	
Sapphire (Al_2O_3)	400 at 25 °C, 275 at 500 °C, 345 at 1000 °C	1900	3.9-4.1
Boron nitride nanotube	N/A	33000	2.62
Diamond	1600	2800	3.5
Graphene	N/A	130000	1.0
First carbon nanotube ropes	?	3600	1.3
Colossal carbon tube	N/A	7000	0.116

Carbon nanotube	N/A	11000-63000	0.037-1.34
Carbon nanotube composites	N/A	1200	N/A
High-strength carbon nanotube film	N/A	9600	N/A
Iron (pure mono-crystal)		3	7.874
Limpet *Patella vulgata* teeth (Goethite)		4900 3000-6500	

a. Many of the values depend on manufacturing process and purity or composition.

b. Multiwalled carbon nanotubes have the highest tensile strength of any material yet measured, with labs producing them at a tensile strength of 63 GPa, still well below their theoretical limit of 300 GPa. The first nanotube ropes (20 mm in length) whose tensile strength was published (in 2000) had a strength of 3.6 GPa. The density depends on the manufacturing method, and the lowest value is 0.037 or 0.55 (solid).

c. The strength of spider silk is highly variable. It depends on many factors including kind of silk (Every spider can produce several for sundry purposes.), species, age of silk, temperature, humidity, swiftness at which stress is applied during testing, length stress is applied, and way the silk is gathered (forced silking or natural spinning). The value shown in the table, 1000 MPa, is roughly representative of the results from a few studies involving several different species of spider however specific results varied greatly.

d. Human hair strength varies by ethnicity and chemical treatments.

Typical properties for annealed elements			
Element	**Young's modulus (GPa)**	**Offset or yield strength (MPa)**	**Ultimate strength (MPa)**
silicon	107		5000–9000
tungsten	411	550	550–620
iron	211	80–100	350
titanium	120	100–225	246–370
copper	130	117	210
tantalum	186	180	200
tin	47	9–14	15–200
zinc alloy	85–105	200–400	200–400
nickel	170	140–350	140–195
silver	83		170
gold	79		100

aluminium	70	15–20	40-50
lead	16		12

Tensile Properties

Stress and Strain

Stress, $s = P/A$

where P is the applied load and A is the original area of the cross section of the sample

$$\text{Strain, } e = \frac{\Delta L}{L_o} = \frac{L - L_o}{L_o}$$

$L_o = Original\,length$
$L = Final\,length$
$\Delta L = L - L_o\ is\,the\,elongation$

These are called engineering stress and engineering strain.

Elastic and Plastic Behavior

- All materials deform when subjected to an external load.
- Up to a certain load the material will recover its original dimensions when the load is released. This is known as elastic behavior.
- The load up to which the materials remains elastic is the elastic limit. The deformation or strain produced within the elastic limit is proportional to the load or stress. This is known as Hook's Law , Stress $\propto$ Strain or Stress = E*Strain. E is known as the Elastic Modulus.
- When the load exceeds the elastic limit, the deformation produced is permanent. This is called plastic deformation. Hook's law is no longer valid in the plastic region.

Tensile Testing

- Load is applied uniaxially in a tensile testing frame and the displacement is recorded.
- The stress and strain are derived using earlier equations
- The stress is plotted against strain to generate the stress- strain curve.
- Different properties are calculated from this curve.

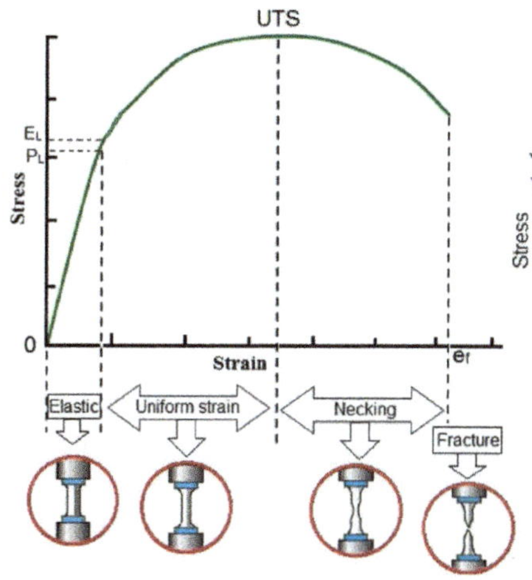

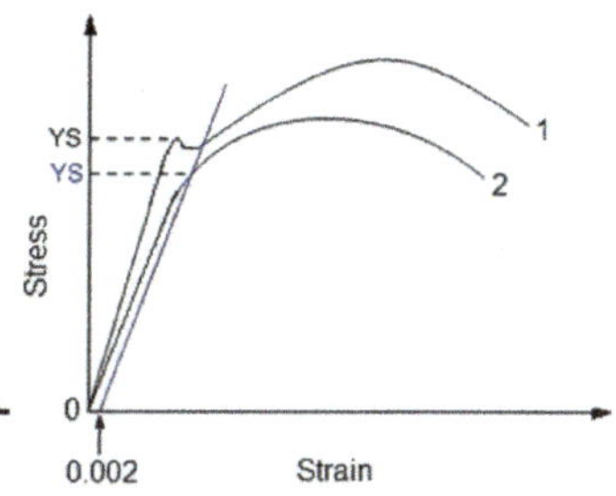

Tensile Properties

- E_L = Elastic limit, up to which Hook's Law (Stress ∝ Strain) is valid. The material comes back to original shape when the load is released.
- Elastic limit is difficult to determine. The proportional limit, P_L, the load at which the curve deviates from linearity, is taken as the elastic portion.
- The slope of the linear region is the Young's Modulus or Elastic Modulus (E).
- Loading beyond P_L produces permanent or plastic deformation. The onset point of plastic deformation is known as Yield stress (YS).
- In some materials like mild steel the yield point is prominent.
- In many other metals and alloys the yield point is not distinct. In such cases, a line parallel to the linear region is drawn at a strain = 0.002 (0.2%) and its intercept on the plastic region is taken as the yield stress (Fig. b). This is called 0.2% Proof stress.
- The stress at the maximum load is called ultimate tensile strength (UTS).
- The strain up to UTS is the uniform plastic strain. Beyond this the cross sectional area reduces and necking takes place.
- The fracture strain $e_f = (L_f - L_o)/L_o$, where L_f is the length after fracture, is taken as the measure of Ductility.

Resilience

The ability of a material to absorb energy in the elastic region. This is given by the strain energy per unit volume $U_o = \frac{1}{2} s_o e_o (= s_o{}^2/2E)$ which is the area of the elastic region

Toughness:

Ability to absorb energy in the plastic range. This is given by the total area under the stress-strain curve

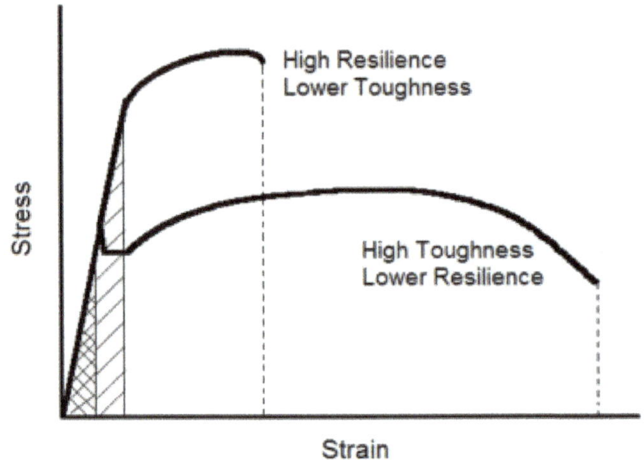

High resilience is a property required in spring steels whereas structural steels have high toughness but lower resilience.

Ductile vs. Brittle Fracture

- The fracture strain $e_f = (L_f - L_o) / L_o$, or reduction of area at fracture, $q = (A - A_o)/A_o$, is taken as the measure of Ductility.
- A ductile material exhibits high fracture strain, that is, it undergoes significant plastic deformation before fracture.
- A brittle material is the one which exhibits little or no plastic deformation before fracture.

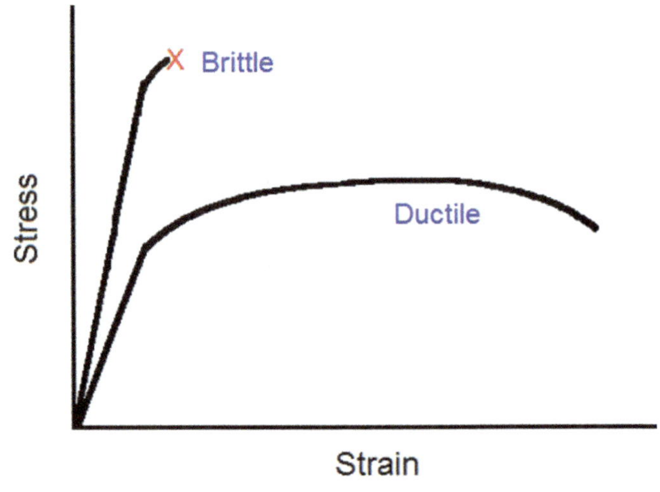

True Stress and Strain

- The engineering stress and strain are based on the original sample dimensions which change during the test.
- True stress and strain on the other hand are based on the actual or instantaneous dimensions and hence, are better representation of the deformation behavior of the material.

True strain $\varepsilon = \sum \frac{L_1 - L_O}{L_O} + \frac{L_2 - L_1}{L_1} + \frac{L_3 - L_2}{L_2}$ ....

$$\varepsilon = \int_{L_O}^{L} \frac{dL}{L} = ln \frac{L}{L_O} = ln\ (e+1$$

Engineering stress, $s = P/A_O$

True stress, $\sigma = P/A = (P/A_O)(A_O/A) = s\,(A_O/A)$

Volume, AL, remains constant, $A_O\ L_O\ = AL \rightarrow A_O/A = L/\ L_O\ = (e+1)$

$$\sigma = s\,(e+1)$$

True Stress-Strain Curve

Since the engineering stress-strain curve is based on original area, it descends after maximum load as the load bearing ability of the sample decreases due to reduction in area.

- The true stress-strain curve (blue) however, continues to go up till fracture as it is based on the actual area.

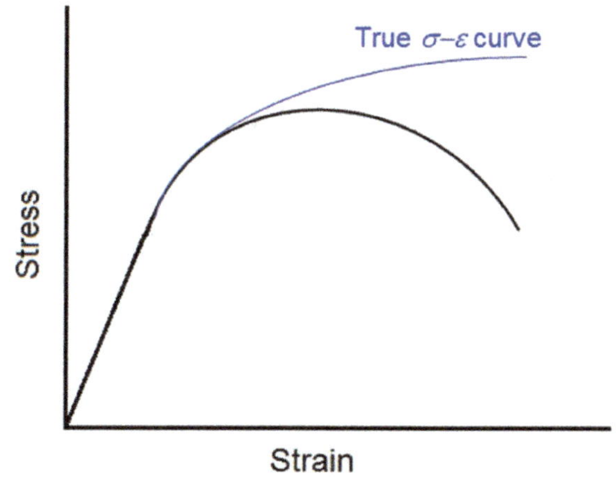

The Flow Curve

The true stress-strain curve is also known as flow curve. The plastic region of the flow curve can be described as

$$\sigma = K\varepsilon^n$$

n is known as strain-hardening exponent and K is the strength coefficient. A log-log plot up to maximum load will yield a straight line. The slope of the line is n. K is the true stress at $\varepsilon = 1$. $n = 0$, perfectly plastic solid, $n = 1$, elastic solid. For most metals $n = 0.1 - 0.5$

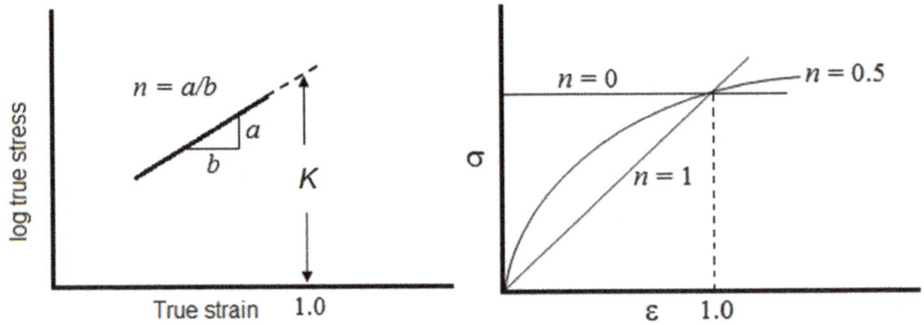

Poisson's Ratio

A tensile force in the x direction produces an extension along that axis while it produces contraction along the transverse y and z axis.

The ratio of the lateral to axial strain is the Poisson's ratio, v. For most metals it is around 0.33

$$V = -\frac{\varepsilon_y}{\varepsilon_x} = -\frac{\varepsilon_z}{\varepsilon_x}$$

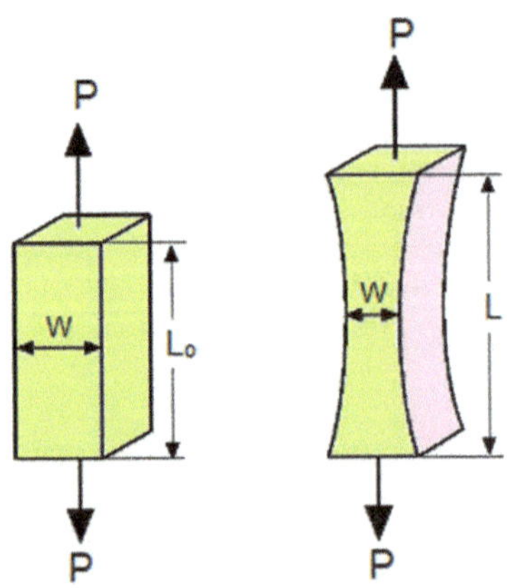

Shear Stress and Strain

The deformation in a body may also result in change in the initial angle between any two lines.

The angular change is known as the shear strain, γ, which is produced by a shear stress, τ.

$$\gamma = \frac{a}{h} = \tan\theta = \theta \qquad \tau = G\gamma$$

G *is the shear* modulus

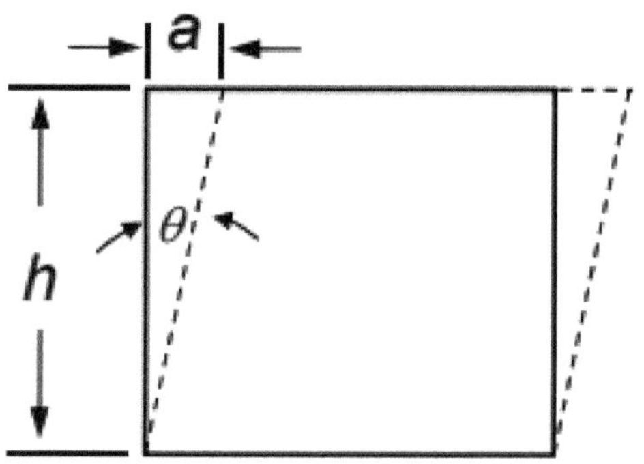

Structure-property Correlation

- Structure-insensitive: Elastic modulus
- Structure-sensitive properties: Yield stress, UTS, Ductility. These properties vary with the structure of the material.
- For example, the same material having a finer grain size will have higher strength as per the relation-

$$\sigma_o = \sigma_i + kd^{-1/2}$$

  This is known as the Hall-Petch equation which relates the yield strength to grain size.

  σ_0 is the yield strength, d is the grain size and σ_i and k are material dependent constants.

- Finer grain size – large grain boundary area/unit volume. As grain boundaries hinder dislocation motion, stress required to move the dislocations increases in the fine grained material and hence the strength increases.

Shear Strength

In engineering, shear strength is the strength of a material or component against the type of yield or structural failure where the material or component fails in shear. A shear load is a force that tends to produce a sliding failure on a material along a plane that is parallel to the direction of the force. When a paper is cut with scissors, the paper fails in shear.

In structural and mechanical engineering, the shear strength of a component is important for designing the dimensions and materials to be used for the manufacture or construction of the component (e.g. beams, plates, or bolts). In a reinforced concrete beam, the main purpose of reinforcing bar (rebar) stirrups is to increase the shear strength.

For shear stress τ applies

$$\tau = \frac{\sigma_1 - \sigma_3}{2},$$

where

σ_1 is major principal stress and

σ_3 is minor principal stress.

In general: ductile materials (e.g. aluminium) fail in shear, whereas brittle materials (e.g. cast iron) fail in tension.

To calculate:

Given total force at failure (F) and the force-resisting area (e.g. the cross-section of a bolt loaded in shear), ultimate shear strength (τ) is:

$$\tau = \frac{F}{A} = \frac{F}{\pi r_{bolt}^2} = \frac{4F}{\pi d_{bolt}^2}$$

Comparison

As a very rough guide relating tensile, yield, and shear strengths:

Material	Ultimate Strength Relationship	Yield Strength Relationship
Steels	USS = approx. 0.75*UTS	SYS = approx. 0.58*TYS
Ductile Iron	USS = approx. 0.9*UTS	SYS = approx. 0.75*TYS
Malleable Iron	USS = approx. 1.0*UTS	
Wrought Iron	USS = approx. 0.83*UTS	
Cast Iron	USS = approx. 1.3*UTS	
Aluminiums	USS = approx. 0.65*UTS	SYS = approx. 0.55*TYS

USS: Ultimate Shear Strength, UTS: Ultimate Tensile Strength, SYS: Shear Yield Stress, TYS: Tensile Yield Stress

Material	Ultimate stress (Ksi)	Ultimate stress (MPa)
Fiberglass/epoxy *(23 ° C)*	7.82	53.9

When values measured from physical samples are desired, a number of testing standards are available, covering different material categories and testing conditions. In the

US, ASTM standards for measuring shear strength include ASTM B831, D732, D4255, D5379, and D7078. Internationally, ISO testing standards for shear strength include ISO 3597, 12579, and 14130.

Shear Modulus

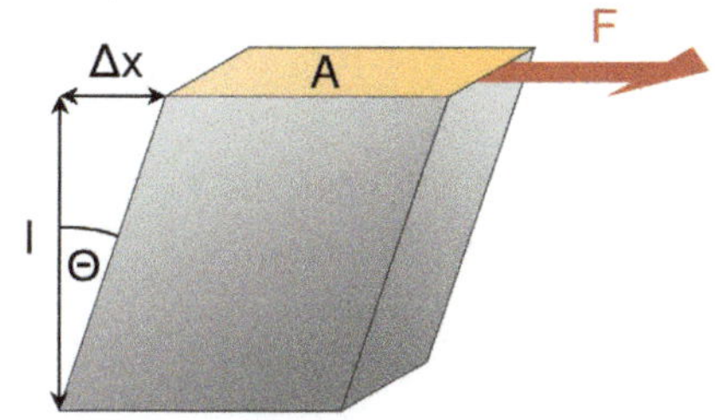

Shear strain

In materials science, shear modulus or modulus of rigidity, denoted by G, or sometimes S or μ, is defined as the ratio of shear stress to the shear strain:

$$G \stackrel{\text{def}}{=} \frac{\tau_{xy}}{\gamma_{xy}} = \frac{F/A}{\Delta x/l} = \frac{Fl}{A\Delta x}$$

where

$\tau_{xy} = F/A =$ shear stress

F is the force which acts

A is the area on which the force acts

$\gamma_{xy} =$ shear strain. In engineering $:= \Delta x/l = \tan\theta$, elsewhere $:= \theta$

Δx is the transverse displacement

l is the initial length

Shear modulus' derived SI unit is the pascal (Pa), although it is usually expressed in gigapascals (GPa) or in thousands of pounds per square inch (ksi). Its dimensional form is $M^1L^{-1}T^{-2}$, replacing *force* by *mass* times *acceleration*.

The shear modulus is always positive.

Explanation

Material	Typical values for shear modulus (GPa) (at room temperature)
Diamond	478.0
Steel	79.3
Copper	44.7

Titanium	41.4
Glass	26.2
Aluminium	25.5
Iron	52.5
Polyethylene	0.117
Rubber	0.0006

The shear modulus is one of several quantities for measuring the stiffness of materials. All of them arise in the generalized Hooke's law:

- Young's modulus E describes the material's strain response to uniaxial stress in the direction of this stress (like pulling on the ends of a wire or putting a weight on top of a column, with the wire getting longer and the column losing height),
- the Poisson's ratio ν describes the response in the directions orthogonal to this uniaxial stress (the wire getting thinner and the column thicker),
- the bulk modulus K describes the material's response to (uniform) hydrostatic pressure (like the pressure at the bottom of the ocean or a deep swimming pool),
- the shear modulus G describes the material's response to shear stress (like cutting it with dull scissors).
- These moduli are not independent, and for isotropic materials they are connected via the equations $2G(1+\nu) = E = 3K(1-2\nu)$.

The shear modulus is concerned with the deformation of a solid when it experiences a force parallel to one of its surfaces while its opposite face experiences an opposing force (such as friction). In the case of an object that's shaped like a rectangular prism, it will deform into a parallelepiped. Anisotropic materials such as wood, paper and also essentially all single crystals exhibit differing material response to stress or strain when tested in different directions. In this case one may need to use the full tensor-expression of the elastic constants, rather than a single scalar value.

One possible definition of a fluid would be a material with zero shear modulus.

Waves

In homogeneous and isotropic solids, there are two kinds of waves, pressure waves and shear waves. The velocity of a shear wave, (v_s) is controlled by the shear modulus,

$$v_s = \sqrt{\frac{G}{\rho}}$$

where

G is the shear modulus

ρ is the solid's density.

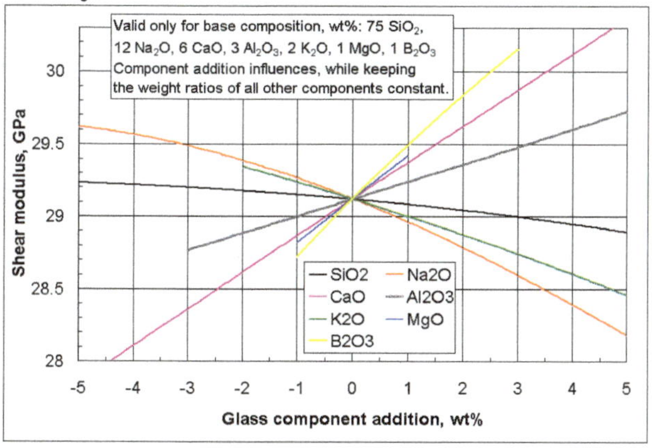

Influences of selected glass component additions on the shear modulus of a specific base glass.

Shear Modulus of Metals

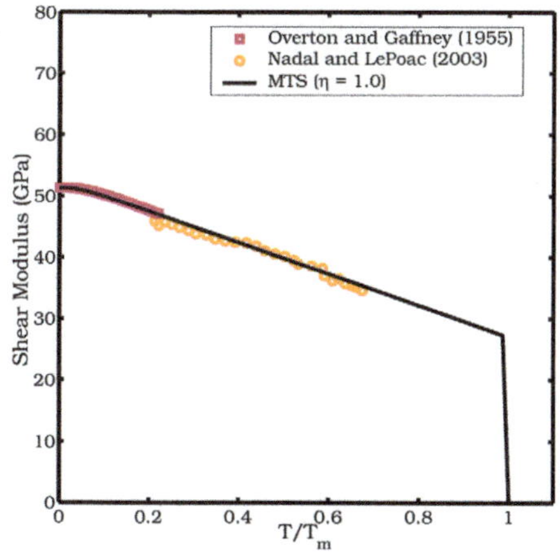

Shear modulus of copper as a function of temperature. The experimental data are shown with colored symbols.

The shear modulus of metals is usually observed to decrease with increasing temperature. At high pressures, the shear modulus also appears to increase with the applied pressure. Correlations between the melting temperature, vacancy formation energy, and the shear modulus have been observed in many metals.

Several models exist that attempt to predict the shear modulus of metals (and possibly that of alloys). Shear modulus models that have been used in plastic flow computations include:

1. the MTS shear modulus model developed by and used in conjunction with the Mechanical Threshold Stress (MTS) plastic flow stress model.
2. the Steinberg-Cochran-Guinan (SCG) shear modulus model developed by and

used in conjunction with the Steinberg-Cochran-Guinan-Lund (SCGL) flow stress model.

3. the Nadal and LePoac (NP) shear modulus model that uses Lindemann theory to determine the temperature dependence and the SCG model for pressure dependence of the shear modulus.

MTS Shear Modulus Model

The MTS shear modulus model has the form:

$$\mu(T) = \mu_0 - \frac{D}{\exp(T_0 / T) - 1}$$

where μ_0 is the shear modulus at $T = 0K$, and D and T_0 are material constants.

SCG Shear Modulus Model

The Steinberg-Cochran-Guinan (SCG) shear modulus model is pressure dependent and has the form

$$\mu(p,T) = \mu_0 + \frac{\partial \mu}{\partial p}\frac{p}{\eta^{1/3}} + \frac{\partial \mu}{\partial T}(T - 300); \quad \eta := \rho / \rho_0$$

where, μ_0 is the shear modulus at the reference state (T = 300 K, p = 0, η = 1), p is the pressure, and T is the temperature.

NP Shear Modulus Model

The Nadal-Le Poac (NP) shear modulus model is a modified version of the SCG model. The empirical temperature dependence of the shear modulus in the SCG model is replaced with an equation based on Lindemann melting theory. The NP shear modulus model has the form:

$$\mu(p,T) = \frac{1}{\mathcal{J}(\hat{T})}\left[\left(\mu_0 + \frac{\partial \mu}{\partial p}\frac{p}{\eta^{1/3}}\right)(1 - \hat{T}) + \frac{\rho}{Cm} k_b\, T\right]; \quad C := \frac{(6\pi^2)^{2/3}}{3} f^2$$

where

$$\mathcal{J}(\hat{T}) := 1 + \exp\left[-\frac{1 + 1/\zeta}{1 + \zeta / (1 - \hat{T})}\right] \quad \text{for} \quad \hat{T} := \frac{T}{T_m} \in [0, 1 + \zeta],$$

and μ_0 is the shear modulus at 0 K and ambient pressure, ζ is a material parameter, k_b is the Boltzmann constant, m is the atomic mass, and f is the Lindemann constant.

Heat Capacity

Heat capacity or thermal capacity is a measurable physical quantity equal to the ratio of the heat added to (or removed from) an object to the resulting temperature change. The unit of heat capacity is joule per kelvin $\frac{J}{K}$, or kilogram metre squared per kelvin second squared $\frac{kgm^2}{Ks^2}$ in the International System of Units (SI). The dimensional form is $L^2MT^{-2}\Theta^{-1}$. Specific heat is the amount of heat needed to raise the temperature of one kilogram of mass by 1 kelvin.

Heat capacity is an extensive property of matter, meaning it is proportional to the size of the system. When expressing the same phenomenon as an intensive property, the heat capacity is divided by the amount of substance, mass, or volume, thus the quantity is independent of the size or extent of the sample. The molar heat capacity is the heat capacity per unit amount (SI unit: mole) of a pure substance and the specific heat capacity, often called simply specific heat, is the heat capacity per unit mass of a material. Nonetheless some authors use the term specific heat to refer to the ratio of the specific heat capacity of a substance at any given temperature, to the specific heat capacity of another substance at a reference temperature, much in the fashion of specific gravity. In some engineering contexts, the volumetric heat capacity is used.

Temperature reflects the average randomized kinetic energy of constituent particles of matter (i.e., atoms or molecules) relative to the centre of mass of the system, while heat is the transfer of energy across a system boundary into the body other than by work or matter transfer. Translation, rotation, and vibration of atoms represent the degrees of freedom of motion which classically contribute to the heat capacity of gases, while only vibrations are needed to describe the heat capacities of most solids, as shown by the Dulong–Petit law. Other contributions can come from magnetic and electronic degrees of freedom in solids, but these rarely make substantial contributions.

For quantum mechanical reasons, at any given temperature, some of these degrees of freedom may be unavailable, or only partially available, to store thermal energy. In such cases, the heat capacity is a fraction of the maximum. As the temperature approaches absolute zero, the heat capacity of a system approaches zero, because of loss of available degrees of freedom. Quantum theory can be used to quantitatively predict the heat capacity of simple systems.

History

In a previous theory of heat common in the early modern period, heat was thought to be a measurement of an invisible fluid, known as the *caloric*. Bodies were capable of holding a certain amount of this fluid, hence the term *heat capacity*, named and first investigated by Scottish chemist Joseph Black in the 1750s.

Since the development of thermodynamics in the 18th and 19th centuries, scientists have abandoned the idea of a physical caloric, and instead understand heat as a manifestation of a system's internal energy. Heat is no longer considered a fluid, but rather a transfer of disordered energy. Nevertheless, at least in English, the term "heat capacity" survives. In some other languages, the term *thermal capacity* is preferred, and it is also sometimes used in English.

Units

Extensive Properties

In the International System of Units, heat capacity has the unit joules per kelvin (J/K). The heat capacity (symbol *C*) of a system is defined as the ratio of heat transferred to or from the system and the resulting change in temperature in the system,

$$C(T) = \frac{\delta Q}{dT},$$

where the symbol δ designates heat as a path function. If the temperature change is sufficiently small the heat capacity may be assumed to be constant:

$$C = \frac{Q}{\Delta T}.$$

Heat capacity is an extensive property, meaning it depends on the extent or size of the physical system studied. A sample containing twice the amount of substance as another sample requires the transfer of twice the amount of heat (Q) to achieve the same change in temperature (ΔT).

Intensive Properties

For many purposes it is more convenient to report heat capacity as an intensive property, an intrinsic characteristic of a particular substance. In practice, this is most often an expression of the property in relation to a unit of mass; in science and engineering, such properties are often prefixed with the term *specific*. International standards now recommend that specific heat capacity always refer to division by mass. The units for the specific heat capacity are $[c] = \frac{\mathrm{J}}{\mathrm{kg \cdot K}}$.

In chemistry, heat capacity is often specified relative to one mole, the unit of amount of substance, and is called the molar heat capacity. It has the unit $[C_{\mathrm{mol}}] = \frac{\mathrm{J}}{\mathrm{mol \cdot K}}$.

For some considerations it is useful to specify the volume-specific heat capacity, commonly called volumetric heat capacity, which is the heat capacity per unit volume and has SI units $[s] = \frac{\mathrm{J}}{\mathrm{m^3 \cdot K}}$. This is used almost exclusively for liquids and solids, since for gases it may be confused with specific heat capacity *at constant volume*.

Alternative Unit Systems

While SI units are the most widely used, some countries and industries also use other systems of measurement. One older unit of heat is the kilogram-calorie (Cal), originally defined as the energy required to raise the temperature of one kilogram of water by one degree Celsius, typically from 14.5 to 15.5 °C. The specific average heat capacity of water on this scale would therefore be exactly 1 Cal/(C°·kg). However, due to the temperature-dependence of the specific heat, a large number of different definitions of the calorie came into being. Whilst once it was very prevalent, especially its smaller cgs variant the gram-calorie (cal), defined thus the specific heat of water would be 1 cal/(K·g), in most fields the use of the calorie is now archaic.

In the United States other units of measure for heat capacity may be quoted in disciplines such as construction, civil engineering, and chemical engineering. A still common system is the English Engineering Units in which the mass reference is pound mass and the temperature is specified in degrees Fahrenheit or Rankine. One (rare) unit of heat is the pound calorie (lb-cal), defined as the amount of heat required to raise the temperature of one pound of water by one degree Celsius. On this scale the specific heat of water would be 1 lb-cal/(K·lb). More common is the British thermal unit, the standard unit of heat in the U.S. construction industry. This is defined such that the specific heat of water is 1 BTU/(F°·lb). The path integral Monte Carlo method is a numerical approach for determining the values of heat capacity, based on quantum dynamical principles. However, good approximations can be made for gases in many states using simpler methods outlined below. For many solids composed of relatively heavy atoms (atomic number > iron), at non-cryogenic temperatures, the heat capacity at room temperature approaches 3R = 24.94 joules per kelvin per mole of atoms (Dulong–Petit law, R is the gas constant). Low temperature approximations for both gases and solids at temperatures less than their characteristic Einstein temperatures or Debye temperatures can be made by the methods of Einstein and Debye discussed below. Water (liquid): CP = 4185.5 J/(kg·K) (15 °C, 101.325 kPa) Water (liquid): CVH = 74.539 J/(mol·K) (25 °C) For liquids and gases, it is important to know the pressure to which given heat capacity data refer. Most published data are given for standard pressure. However, different standard conditions for temperature and pressure have been defined by different organizations. The International Union of Pure and Applied Chemistry (IUPAC) changed its recommendation from one atmosphere to the round value 100 kPa (≈750.062 Torr).

Measurement

It may appear that the way to measure heat capacity is to add a known amount of heat to an object, and measure the change in temperature. This works reasonably well for many solids. However, for precise measurements, and especially for gases, other aspects of measurement become critical.

The heat capacity can be affected by many of the state variables that describe the thermodynamic system under study. These include the starting and ending temperature, as well as the pressure and the volume of the system before and after heat is added. So rather than a single way to measure heat capacity, there are actually several slightly different measurements of heat capacity. The most commonly used methods for measurement are to hold the object either at constant pressure (C_P) or at constant volume (C_V). Gases and liquids are typically also measured at constant volume. Measurements under constant pressure produce larger values than those at constant volume because the constant pressure values also include heat energy that is used to do work to expand the substance against the constant pressure as its temperature increases. This difference is particularly notable in gases where values under constant pressure are typically 30% to 66.7% greater than those at constant volume. Hence the heat capacity ratio of gases is typically between 1.3 and 1.67.

The specific heat capacities of substances comprising molecules (as distinct from monatomic gases) are not fixed constants and vary somewhat depending on temperature. Accordingly, the temperature at which the measurement is made is usually also specified. Examples of two common ways to cite the specific heat of a substance are as follows:

- Water (liquid): $C_P = 4185.5$ J/(kg·K) (15 °C, 101.325 kPa)
- Water (liquid): $C_V H = 74.539$ J/(mol·K) (25 °C)

For liquids and gases, it is important to know the pressure to which given heat capacity data refer. Most published data are given for standard pressure. However, quite different standard conditions for temperature and pressure have been defined by different organizations. The International Union of Pure and Applied Chemistry (IUPAC) changed its recommendation from one atmosphere to the round value 100 kPa (≈750.062 Torr).

Calculation from First Principles

The path integral Monte Carlo method is a numerical approach for determining the values of heat capacity, based on quantum dynamical principles. However, good approximations can be made for gases in many states using simpler methods outlined below. For many solids composed of relatively heavy atoms (atomic number > iron), at non-cryogenic temperatures, the heat capacity at room temperature approaches $3R = 24.94$ joules per kelvin per mole of atoms (Dulong–Petit law, R is the gas constant). Low temperature approximations for both gases and solids at temperatures less than their characteristic Einstein temperatures or Debye temperatures can be made by the methods of Einstein and Debye discussed below.

Thermodynamic Relations and Definition of Heat Capacity

The internal energy of a closed system changes either by adding heat to the system or by the system performing work. Written mathematically we have

$$\Delta e_{\text{system}} = e_{\text{in}} - e_{\text{out}}$$

Or

$$\mathrm{d}U = \delta Q + \delta W.$$

For work as a result of an increase of the system volume we may write,

$$\mathrm{d}U = \delta Q - P\mathrm{d}V.$$

If the heat is added at constant volume, then the second term of this relation vanishes and one readily obtains

$$\left(\frac{\partial U}{\partial T}\right)_{\mathrm{V}} = \left(\frac{\partial Q}{\partial T}\right)_{\mathrm{V}} = C_{\mathrm{V}}.$$

This defines the *heat capacity at constant volume*, C_V, which is also related to changes in internal energy. Another useful quantity is the *heat capacity at constant pressure*, C_P. This quantity refers to the change in the *enthalpy* of the system, which is given by

$$H = U + PV.$$

A small change in the enthalpy can be expressed as

$$\mathrm{d}H = \delta Q + V\mathrm{d}P,$$

and therefore, at constant pressure, we have

$$\left(\frac{\partial H}{\partial T}\right)_{\mathrm{P}} = \left(\frac{\partial Q}{\partial T}\right)_{\mathrm{P}} = C_{\mathrm{P}}.$$

These two equations:

$$\left(\frac{\partial U}{\partial T}\right)_{\mathrm{V}} = \left(\frac{\partial Q}{\partial T}\right)_{\mathrm{V}} = C_{\mathrm{V}}.$$

$$\left(\frac{\partial H}{\partial T}\right)_{\mathrm{P}} = \left(\frac{\partial Q}{\partial T}\right)_{\mathrm{P}} = C_{\mathrm{P}}.$$

are property relations and are therefore independent of the type of process. In other words, they are valid for any substance going through any process. Both the internal energy and enthalpy of a substance can change with the transfer of energy in many forms i.e., heat.

Relation between Heat Capacities

Measuring the heat capacity, sometimes referred to as specific heat, at constant volume can be prohibitively difficult for liquids and solids. That is, small temperature changes typically require large pressures to maintain a liquid or solid at constant volume implying the containing vessel must be nearly rigid or at least very strong. Instead it is easier to measure the heat capacity at constant pressure (allowing the material to expand or contract freely) and solve for the heat capacity at constant volume using mathematical relationships derived from the basic thermodynamic laws. Starting from the fundamental thermodynamic relation one can show

$$C_P - C_V = T\left(\frac{\partial P}{\partial T}\right)_{V,n}\left(\frac{\partial V}{\partial T}\right)_{P,n}$$

where the partial derivatives are taken at constant volume and constant number of particles, and constant pressure and constant number of particles, respectively.

This can also be rewritten

$$C_P - C_V = VT\frac{\alpha^2}{\beta_T}$$

where

α is the coefficient of thermal expansion,

β_T is the isothermal compressibility.

The heat capacity ratio or adiabatic index is the ratio of the heat capacity at constant pressure to heat capacity at constant volume. It is sometimes also known as the isentropic expansion factor.

Ideal Gas

For an ideal gas, evaluating the partial derivatives above according to the equation of state where R is the gas constant for an ideal gas

$$PV = nRT$$

$$C_P - C_V = T\left(\frac{\partial P}{\partial T}\right)_{V,n}\left(\frac{\partial V}{\partial T}\right)_{P,n}$$

$$P = \frac{nRT}{V} \Rightarrow \left(\frac{\partial P}{\partial T}\right)_{V,n} = \frac{nR}{V}$$

$$V = \frac{nRT}{P} \Rightarrow \left(\frac{\partial V}{\partial T}\right)_{\mathrm{P,n}} = \frac{nR}{P}$$

substituting

$$T\left(\frac{\partial P}{\partial T}\right)_{\mathrm{V,n}}\left(\frac{\partial V}{\partial T}\right)_{\mathrm{P,n}} = T\left(\frac{nR}{V}\right)\left(\frac{nR}{P}\right) = \left(\frac{nRT}{V}\right)\left(\frac{nR}{P}\right) = P\left(\frac{nR}{P}\right) = nR$$

this equation reduces simply to Mayer's relation,

$$C_{\mathrm{P,m}} - C_{\mathrm{V,m}} = R$$

Specific Heat Capacity

The specific heat capacity of a material on a per mass basis is

$$c = \frac{\partial C}{\partial m},$$

which in the absence of phase transitions is equivalent to

$$c = E_{\mathrm{m}} = \frac{C}{m} = \frac{C}{\rho V},$$

where

C is the heat capacity of a body made of the material in question,

m is the mass of the body,

V is the volume of the body, and

$\rho = \frac{m}{V}$ is the density of the material.

For gases, and also for other materials under high pressures, there is need to distinguish between different boundary conditions for the processes under consideration (since values differ significantly between different conditions). Typical processes for which a heat capacity may be defined include isobaric (constant pressure, $\mathrm{d}P = 0$) or isochoric (constant volume, $\mathrm{d}V = 0$) processes. The corresponding specific heat capacities are expressed as

$$c_{\mathrm{P}} = \left(\frac{\partial C}{\partial m}\right)_{\mathrm{P}},$$

$$c_{\mathrm{V}} = \left(\frac{\partial C}{\partial m}\right)_{\mathrm{V}}.$$

From the dividing through by the mass gives the relation

$$c_P - c_V = \frac{\alpha^2 T}{\rho \beta_T}.$$

A related parameter to c is CV^{-1}, the volumetric heat capacity. In engineering practice, c_V for solids or liquids often signifies a volumetric heat capacity, rather than a constant-volume one. In such cases, the mass-specific heat capacity (specific heat) is often explicitly written with the subscript m, as c_m. Of course, from the above relationships, for solids one writes

$$c_m = \frac{C}{m} = \frac{c_{volumetric}}{\rho}.$$

For pure homogeneous chemical compounds with established molecular or molar mass, or a molar quantity, heat capacity as an intensive property can be expressed on a per mole basis instead of a per mass basis by the following equations analogous to the per mass equations:

$$C_{P,m} = \left(\frac{\partial C}{\partial n}\right)_P = \text{molar heat capacity at constant pressure}$$

$$C_{V,m} = \left(\frac{\partial C}{\partial n}\right)_V = \text{molar heat capacity at constant volume}$$

where n is the number of moles in the body or thermodynamic system. One may refer to such a *per mole* quantity as molar heat capacity to distinguish it from specific heat capacity on a per mass basis.

Polytropic Heat Capacity

The polytropic heat capacity is calculated at processes if all the thermodynamic properties (pressure, volume, temperature) change

$$C_{i,m} = \left(\frac{\partial C}{\partial n}\right) = \text{molar heat capacity at polytropic process}$$

The most important polytropic processes run between the adiabatic and the isotherm functions, the polytropic index is between 1 and the adiabatic exponent (γ or κ)

Dimensionless Heat Capacity

The dimensionless heat capacity of a material is

$$C^* = \frac{C}{nR} = \frac{C}{Nk}$$

where

C is the heat capacity of a body made of the material in question (J/K)

n is the amount of substance in the body (mol)

R is the gas constant (J/(K·mol))

N is the number of molecules in the body. (dimensionless)

k is Boltzmann's constant (J/(K·molecule))

In the ideal gas article, dimensionless heat capacity C^* is expressed as $\hat{c}$, and is related there directly to half the number of degrees of freedom per particle. This holds true for quadratic degrees of freedom, a consequence of the equipartition theorem.

More generally, the dimensionless heat capacity relates the logarithmic increase in temperature to the increase in the dimensionless entropy per particle $S^* = S / Nk$, measured in nats.

$$C^* = \frac{\mathrm{d}S^*}{\mathrm{d}\ln T}\ \mathrm{t}$$

Alternatively, using base 2 logarithms, C^* relates the base-2 logarithmic increase in temperature to the increase in the dimensionless entropy measured in bits.

Heat Capacity at Absolute Zero

From the definition of entropy

$$T\,\mathrm{d}S = \delta Q$$

the absolute entropy can be calculated by integrating from zero kelvins temperature to the final temperature T_f

$$S(T_\mathrm{f}) = \int_{T=0}^{T_\mathrm{f}} \frac{\delta Q}{T} = \int_0^{T_\mathrm{f}} \frac{\delta Q}{\mathrm{d}T}\frac{\mathrm{d}T}{T} = \int_0^{T_\mathrm{f}} C(T)\frac{\mathrm{d}T}{T}.$$

The heat capacity must be zero at zero temperature in order for the above integral not to yield an infinite absolute entropy, which would violate the third law of thermodynamics. One of the strengths of the Debye model is that (unlike the preceding Einstein model) it predicts the proper mathematical form of the approach of heat capacity toward zero, as absolute zero temperature is approached.

Negative Heat Capacity (Stars)

Most physical systems exhibit a positive heat capacity. However, even though it can seem paradoxical at first, there are some systems for which the heat capacity is *nega-*

tive. These are inhomogeneous systems which do not meet the strict definition of thermodynamic equilibrium. They include gravitating objects such as stars, galaxies; and also sometimes some nano-scale clusters of a few tens of atoms, close to a phase transition. A negative heat capacity can result in a negative temperature.

According to the virial theorem, for a self-gravitating body like a star or an interstellar gas cloud, the average potential energy U_{Pot} and the average kinetic energy U_{Kin} are locked together in the relation

$$U_{Pot} = -2U_{Kin},$$

The total energy U (= $U_{Pot} + U_{Kin}$) therefore obeys

$$U = -U_{Kin},$$

If the system loses energy, for example by radiating energy away into space, the average kinetic energy actually increases. If a temperature is defined by the average kinetic energy, then the system therefore can be said to have a negative heat capacity.

A more extreme version of this occurs with black holes. According to black hole thermodynamics, the more mass and energy a black hole absorbs, the colder it becomes. In contrast, if it is a net emitter of energy, through Hawking radiation, it will become hotter and hotter until it boils away.

Theory

Factors that Affect Specific Heat Capacity

For any given substance, the heat capacity of a body is directly proportional to the amount of substance it contains (measured in terms of mass or moles or volume). Doubling the amount of substance in a body doubles its heat capacity, etc.

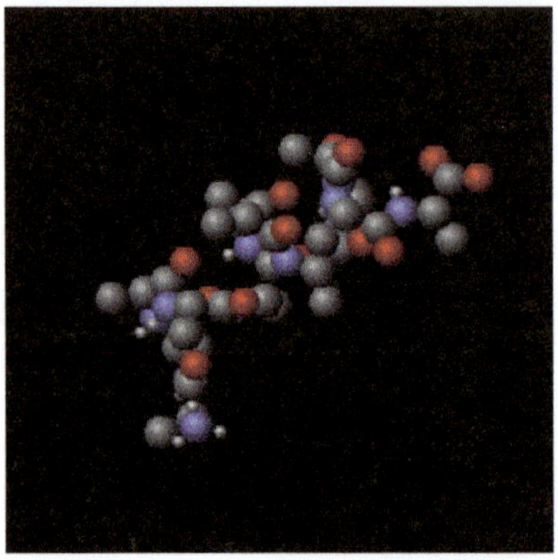

Molecules undergo many characteristic internal vibrations. Potential energy stored in these internal degrees of freedom contributes to a sample's energy content, but not to its temperature. More internal degrees of freedom tend to increase a substance's specific heat capacity, so long as temperatures are high enough to overcome quantum effects.

However, when this effect has been corrected for, by dividing the heat capacity by the quantity of substance in a body, the resulting specific heat capacity is a function of the structure of the substance itself. In particular, it depends on the number of degrees of freedom that are available to the particles in the substance; each independent degree of freedom allows the particles to store thermal energy. The translational kinetic energy of substance particles which manifests as *temperature change* is only one of the many possible degrees of freedom, and thus the larger the number of degrees of freedom available to the particles of a substance *other* than translational kinetic energy, the larger will be the specific heat capacity for the substance. For example, rotational kinetic energy of gas molecules stores heat energy in a way that increases heat capacity, since this energy does not contribute to temperature.

In addition, quantum effects require that whenever energy be stored in any mechanism associated with a bound system which confers a degree of freedom, it must be stored in certain minimal-sized deposits (quanta) of energy, or else not stored at all. Such effects limit the full ability of some degrees of freedom to store energy when their lowest energy storage quantum amount is not easily supplied at the average energy of particles at a given temperature. In general, for this reason, specific heat capacities tend to fall at lower temperatures where the average thermal energy available to each particle degree of freedom is smaller, and thermal energy storage begins to be limited by these quantum effects. Due to this process, as temperature falls toward absolute zero, so also does heat capacity.

Degrees of Freedom

Molecules are quite different from the monatomic gases like helium and argon. With monatomic gases, thermal energy comprises only translational motions. Translational motions are ordinary, whole-body movements in 3D space whereby particles move about and exchange energy in collisions—like rubber balls in a vigorously shaken container. These simple movements in the three dimensions of space mean individual atoms have three translational degrees of freedom. A degree of freedom is any form of energy in which heat transferred into an object can be stored. This can be in translational kinetic energy, rotational kinetic energy, or other forms such as potential energy in vibrational modes. Only three translational degrees of freedom (corresponding to the three independent directions in space) are available for any individual atom, whether it is free, as a monatomic molecule, or bound into a polyatomic molecule.

As to rotation about an atom's axis (again, whether the atom is bound or free), its energy of rotation is proportional to the moment of inertia for the atom, which is extremely small compared to moments of inertia of collections of atoms. This is because almost all of the mass of a single atom is concentrated in its nucleus, which has a radius too small to give a significant moment of inertia. In contrast, the *spacing* of quantum energy levels for a rotating object is inversely proportional to its moment of inertia, and so this spacing becomes very large for objects with very small moments of inertia. For these reasons, the contribution from rotation of atoms on their axes is essentially zero in monatomic gases, because the energy spacing of the associated quantum levels is too large for significant thermal energy to be stored in rotation of systems with such small moments of inertia. For similar reasons, axial rotation around bonds joining atoms in diatomic gases (or along the linear axis in a linear molecule of any length) can also be neglected as a possible "degree of freedom" as well, since such rotation is similar to rotation of monatomic atoms, and so occurs about an axis with a moment of inertia too small to be able to store significant heat energy.

In polyatomic molecules, other rotational modes may become active, due to the much higher moments of inertia about certain axes which do not coincide with the linear axis of a linear molecule. These modes take the place of some translational degrees of freedom for individual atoms, since the atoms are moving in 3-D space, as the molecule rotates. The narrowing of quantum mechanically determined energy spacing between rotational states results from situations where atoms are rotating around an axis that does not connect them, and thus form an assembly that has a large moment of inertia. This small difference between energy states allows the kinetic energy of this type of rotational motion to store heat energy at ambient temperatures. Furthermore, internal vibrational degrees of freedom also may become active (these are also a type of translation, as seen from the view of each atom). In summary, molecules are complex objects with a population of atoms that may move about within the molecule in a number of different ways, and each of these ways of moving is capable of storing energy if the temperature is sufficient.

The heat capacity of molecular substances (on a "per-atom" or atom-molar, basis) does not exceed the heat capacity of monatomic gases, unless vibrational modes are brought into play. The reason for this is that vibrational modes allow energy to be stored as potential energy in inter-atomic bonds in a molecule, which are not available to atoms in monatomic gases. Up to about twice as much energy (on a per-atom basis) per unit of temperature increase can be stored in a solid as in a monatomic gas, by this mechanism of storing energy in the potentials of interatomic bonds. This gives many solids about twice the atom-molar heat capacity at room temperature of monatomic gases.

However, quantum effects heavily affect the actual ratio at lower temperatures (i.e., much lower than the melting temperature of the solid), especially in solids with light and tightly bound atoms (e.g., beryllium metal or diamond). Polyatomic gases store intermediate amounts of energy, giving them a "per-atom" heat capacity that is between that of monatomic gases ($^3/_2$ R per mole of atoms, where R is the ideal gas constant), and the maximum of fully excited warmer solids (3 R per mole of atoms). For gases, heat capacity never falls below the minimum of $^3/_2$ R per mole (of molecules), since the kinetic energy of gas molecules is always available to store at least this much thermal energy. However, at cryogenic temperatures in solids, heat capacity falls toward zero, as temperature approaches absolute zero.

Example of Temperature-dependent Specific Heat Capacity, in a Diatomic Gas

To illustrate the role of various degrees of freedom in storing heat, we may consider nitrogen, a diatomic molecule that has five active degrees of freedom at room temperature: the three comprising translational motions plus two rotational degrees of freedom internally. Although the constant-volume molar heat capacity of nitrogen at this temperature is five-thirds that of monatomic gases, on a per-mole of atoms basis, it is five-sixths that of a monatomic gas. The reason for this is the loss of a degree of freedom due to the bond when it does not allow storage of thermal energy. Two separate nitrogen atoms would have a total of six degrees of freedom—the three translational degrees of freedom of each atom. When the atoms are bonded the molecule will still only have three translational degrees of freedom, as the two atoms in the molecule move as one. However, the molecule cannot be treated as a point object, and the moment of inertia has increased sufficiently about two axes to allow two rotational degrees of freedom to be active at room temperature to give five degrees of freedom. The moment of inertia about the third axis remains small, as this is the axis passing through the centres of the two atoms, and so is similar to the small moment of inertia for atoms of a monatomic gas. Thus, this degree of freedom does not act to store heat, and does not contribute to the heat capacity of nitrogen. The heat capacity *per atom* for nitrogen (5/2 R per mole molecules = 5/4 R per mole atoms) is therefore less than for a monatomic gas (3/2 R per mole molecules or atoms), so long as the temperature remains low enough that no vibrational degrees of freedom are activated.

At higher temperatures, however, nitrogen gas gains one more degree of internal freedom, as the molecule is excited into higher vibrational modes that store thermal energy. A vibrational degree of freedom contributes a heat capacity of 1/2 R each for kinetic and potential energy, for a total of R. Now the bond is contributing heat capacity, and (because of storage of energy in potential energy) is contributing more than if the atoms were not bonded. With full thermal excitation of bond vibration, the heat capacity per volume, or per mole of gas *molecules* approaches seven-thirds that of monatomic gases. Significantly, this is seven-sixths of the monatomic gas value on a mole-of-atoms basis, so this is now a *higher* heat capacity *per atom* than the monatomic figure, because the vibrational mode enables for diatomic gases allows an extra degree of *potential energy* freedom per pair of atoms, which monatomic gases cannot possess.

However, even at these large temperatures where gaseous nitrogen is able to store $7/6^{\text{ths}}$ of the energy *per atom* of a monatomic gas (making it more efficient at storing energy on an atomic basis), it still only stores $7/12^{\text{ths}}$ of the maximal per-atom heat capacity of a *solid*, meaning it is not nearly as efficient at storing thermal energy on an atomic basis, as solid substances can be. This is typical of gases, and results because many of the potential bonds which might be storing potential energy in gaseous nitrogen (as opposed to solid nitrogen) are lacking, because only one of the spatial dimensions for each nitrogen atom offers a bond into which potential energy can be stored without increasing the kinetic energy of the atom. In general, solids are most efficient, on an atomic basis, at storing thermal energy (that is, they have the highest per-atom or per-mole-of-atoms heat capacity).

Per Mole of Different Units

Per Mole of Molecules

When the specific heat capacity, *c*, of a material is measured (lowercase *c* means the unit quantity is in terms of mass), different values arise because different substances have different molar masses (essentially, the weight of the individual atoms or molecules). In solids, thermal energy arises due to the number of atoms that are vibrating. "Molar" heat capacity *per mole of molecules*, for both gases and solids, offer figures which are arbitrarily large, since molecules may be arbitrarily large. Such heat capacities are thus not intensive quantities for this reason, since the quantity of mass being considered can be increased without limit.

Per Mole of Atoms

Conversely, for *molecular-based* substances (which also absorb heat into their internal degrees of freedom), massive, complex molecules with high atomic count—like octane—can store a great deal of energy per mole and yet are quite unremarkable on a mass basis, or on a per-atom basis. This is because, in fully excited systems, heat is stored independently by each atom in a substance, not primarily by the bulk motion of molecules.

Thus, it is the heat capacity per-mole-of-atoms, not per-mole-of-molecules, which is the intensive quantity, and which comes closest to being a constant for all substances at high temperatures. This relationship was noticed empirically in 1819, and is called the Dulong–Petit law, after its two discoverers. Historically, the fact that specific heat capacities are approximately equal when corrected by the presumed weight of the atoms of solids, was an important piece of data in favor of the atomic theory of matter.

Because of the connection of heat capacity to the number of atoms, some care should be taken to specify a mole-of-molecules basis vs. a mole-of-atoms basis, when comparing specific heat capacities of molecular solids and gases. Ideal gases have the same numbers of molecules per volume, so increasing molecular complexity adds heat capacity on a per-volume and per-mole-of-molecules basis, but may lower or raise heat capacity on a per-atom basis, depending on whether the temperature is sufficient to store energy as atomic vibration.

In solids, the quantitative limit of heat capacity in general is about 3 R per mole of atoms, where R is the ideal gas constant. This 3 R value is about 24.9 J/mole.K. Six degrees of freedom (three kinetic and three potential) are available to each atom. Each of these six contributes 1/2 R specific heat capacity per mole of atoms. This limit of 3 R per mole specific heat capacity is approached at room temperature for most solids, with significant departures at this temperature only for solids composed of the lightest atoms which are bound very strongly, such as beryllium (where the value is only of 66% of 3 R), or diamond (where it is only 24% of 3 R). These large departures are due to quantum effects which prevent full distribution of heat into all vibrational modes, when the energy difference between vibrational quantum states is very large compared to the average energy available to each atom from the ambient temperature.

For monatomic gases, the specific heat is only half of 3 R per mole, i.e. (3/2 R per mole) due to loss of all potential energy degrees of freedom in these gases. For polyatomic gases, the heat capacity will be intermediate between these values on a per-mole-of-atoms basis, and (for heat-stable molecules) would approach the limit of 3 R per mole of atoms, for gases composed of complex molecules, and at higher temperatures at which all vibrational modes accept excitational energy. This is because very large and complex gas molecules may be thought of as relatively large blocks of solid matter which have lost only a relatively small fraction of degrees of freedom, as compared to a fully integrated solid.

For a list of heat capacities per atom-mole of various substances, in terms of R, see the last column of the table of heat capacities below.

Corollaries of These Considerations for Solids (Volume-specific Heat Capacity)

Since the bulk density of a solid chemical element is strongly related to its molar mass (usually about 3 R per mole, as noted above), there exists a noticeable inverse correlation between a solid's density and its specific heat capacity on a per-mass basis. This is due to a very approximate tendency of atoms of most elements to be about the same size (and constancy of mole-specific heat capacity) resulting in a good correlation between the *volume* of any given solid chemical element and its total heat capacity. Another way of stating this, is that the volume-specific heat capacity (volumetric heat capacity) of solid elements is roughly a constant. The molar volume of solid elements is very roughly constant, and (even more reliably) so also is the molar heat capacity for most solid substances. These two factors determine the volumetric heat capacity, which as a bulk property may be striking in consistency. For example, the element uranium is a metal which has a density almost 36 times that of the metal lithium, but uranium's specific heat capacity on a volumetric basis (i.e. per given volume of metal) is only 18% larger than lithium's.

Since the volume-specific corollary of the Dulong–Petit specific heat capacity relationship requires that atoms of all elements take up (on average) the same volume in solids, there are many departures from it, with most of these due to variations in atomic size. For instance, arsenic, which is only 14.5% less dense than antimony, has nearly 59% more specific heat capacity on a mass basis. In other words; even though an ingot of arsenic is only about 17% larger than an antimony one of the same mass, it absorbs about 59% more heat for a given temperature rise. The heat capacity ratios of the two substances closely follows the ratios of their molar volumes (the ratios of numbers of atoms in the same volume of each substance); the departure from the correlation to simple volumes in this case is due to lighter arsenic atoms being significantly more closely packed than antimony atoms, instead of similar size. In other words, similar-sized atoms would cause a mole of arsenic to be 63% larger than a mole of antimony, with a correspondingly lower density, allowing its volume to more closely mirror its heat capacity behavior.

Other Factors

Hydrogen Bonds

Hydrogen-containing polar molecules like ethanol, ammonia, and water have powerful, intermolecular hydrogen bonds when in their liquid phase. These bonds provide another place where heat may be stored as potential energy of vibration, even at comparatively low temperatures. Hydrogen bonds account for the fact that liquid water stores nearly the theoretical limit of 3 R per mole of atoms, even at relatively low temperatures (i.e. near the freezing point of water).

Impurities

In the case of alloys, there are several conditions in which small impurity concentrations can greatly affect the specific heat. Alloys may exhibit marked difference in behaviour even in the case of small amounts of impurities being one element of the alloy; for example impurities in semiconducting ferromagnetic alloys may lead to quite different specific heat properties.

The Simple Case of the Monatomic Gas

In the case of a monatomic gas such as helium under constant volume, if it is assumed that no electronic or nuclear quantum excitations occur, each atom in the gas has only 3 degrees of freedom, all of a translational type. No energy dependence is associated with the degrees of freedom which define the position of the atoms. While, in fact, the degrees of freedom corresponding to the momenta of the atoms are quadratic, and thus contribute to the heat capacity. There are N atoms, each of which has 3 components of momentum, which leads to $3N$ total degrees of freedom. This gives:

$$C_V = \left(\frac{\partial U}{\partial T}\right)_V = \frac{3}{2} N k_B = \frac{3}{2} nR$$

$$C_{V,m} = \frac{C_V}{n} = \frac{3}{2} R$$

where

C_V is the heat capacity at constant volume of the gas

$C_{V,m}$ is the *molar heat capacity* at constant volume of the gas

N is the total number of atoms present in the container

n is the number of moles of atoms present in the container (n is the ratio of N and Avogadro's number)

R is the ideal gas constant, (8.3144621 J/(mol·K). R is equal to the product of Boltzmann's constant k_B and Avogadro's number

The following table shows experimental molar constant volume heat capacity measurements taken for each noble monatomic gas (at 1 atm and 25 °C):

Monatomic gas	$C_{V,m}$ (J/(mol·K))	$C_{V,m}/R$
He	12.5	1.50
Ne	12.5	1.50
Ar	12.5	1.50
Kr	12.5	1.50
Xe	12.5	1.50

It is apparent from the table that the experimental heat capacities of the monatomic noble gases agrees with this simple application of statistical mechanics to a very high degree.

The molar heat capacity of a monatomic gas at constant pressure is then

$$C_{p,m} = C_{V,m} + R = \frac{5}{2}R$$

Diatomic Gas

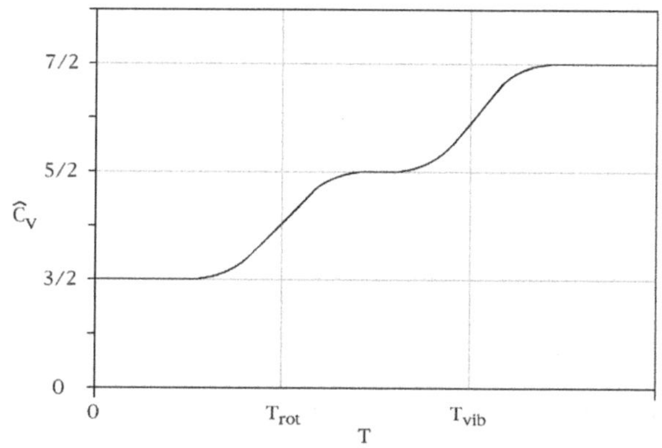

Constant volume specific heat capacity of a diatomic gas (idealised). As temperature increases, heat capacity goes from 3/2 R (translation contribution only), to 5/2 R (translation plus rotation), finally to a maximum of 7/2 R (translation + rotation + vibration)

In the somewhat more complex case of an ideal gas of diatomic molecules, the presence of internal degrees of freedom are apparent. In addition to the three translational degrees of freedom, there are rotational and vibrational degrees of freedom. In general, the number of degrees of freedom, f, in a molecule with n_a atoms is $3n_a$:

$$f = 3n_a$$

Mathematically, there are a total of three rotational degrees of freedom, one corresponding to rotation about each of the axes of three-dimensional space. However, in practice only the existence of two degrees of rotational freedom for linear molecules will be considered. This approximation is valid because the moment of inertia about the internuclear axis is vanishingly small with respect to other moments of inertia in the molecule (this is due to the very small rotational moments of single atoms, due to the concentration of almost all their mass at their centers; compare also the extremely small radii of the atomic nuclei compared to the distance between them in a diatomic molecule). Quantum mechanically, it can be shown that the interval between successive rotational energy eigenstates is inversely proportional to the moment of inertia about that axis. Because the moment of inertia about the internuclear axis is vanishingly small relative to the other two rotational axes, the energy spacing can be considered so high that no excitations of the rotational state can occur unless the temperature is extremely high. It is easy to calculate the expected number of vibrational degrees of freedom (or vibrational modes). There are three degrees of translational freedom, and two degrees of rotational freedom, therefore

$$f_{\text{vib}} = f - f_{\text{trans}} - f_{\text{rot}} = 6 - 3 - 2 = 1$$

Each rotational and translational degree of freedom will contribute $R/2$ in the total molar heat capacity of the gas. Each vibrational mode will contribute R to the total molar heat capacity, however. This is because for each vibrational mode, there is a potential and kinetic energy component. Both the potential and kinetic components will contribute $R/2$ to the total molar heat capacity of the gas. Therefore, a diatomic molecule would be expected to have a molar constant-volume heat capacity of

$$C_{V,m} = \frac{3R}{2} + R + R = \frac{7R}{2} = 3.5R$$

where the terms originate from the translational, rotational, and vibrational degrees of freedom, respectively.

The following is a table of some molar constant-volume heat capacities of various diatomic gases at standard temperature (25 °C = 298 K)

Diatomic gas	$C_{V,m}$ **(J/(mol·K))**	$C_{V,m}/R$
H_2	20.18	2.427
CO	20.2	2.43

N_2	19.9	2.39
Cl_2	24.1	3.06
Br_2 (vapour)	28.2	3.39

From the above table, clearly there is a problem with the above theory. All of the diatomics examined have heat capacities that are lower than those predicted by the equipartition theorem, except Br_2. However, as the atoms composing the molecules become heavier, the heat capacities move closer to their expected values. One of the reasons for this phenomenon is the quantization of vibrational, and to a lesser extent, rotational states. In fact, if it is assumed that the molecules remain in their lowest energy vibrational state because the inter-level energy spacings for vibration-energies are large, the predicted molar constant volume heat capacity for a diatomic molecule becomes just that from the contributions of translation and rotation:

$$C_{V,m} = \frac{3R}{2} + R = \frac{5R}{2} = 2.5R$$

which is a fairly close approximation of the heat capacities of the lighter molecules in the above table. If the quantum harmonic oscillator approximation is made, it turns out that the quantum vibrational energy level spacings are actually inversely proportional to the square root of the reduced mass of the atoms composing the diatomic molecule. Therefore, in the case of the heavier diatomic molecules such as chlorine or bromine, the quantum vibrational energy level spacings become finer, which allows more excitations into higher vibrational levels at lower temperatures. This limit for storing heat capacity in vibrational modes, as discussed above, becomes 7R/2 = 3.5 R per mole of gas molecules, which is fairly consistent with the measured value for Br_2 at room temperature. As temperatures rise, all diatomic gases approach this value.

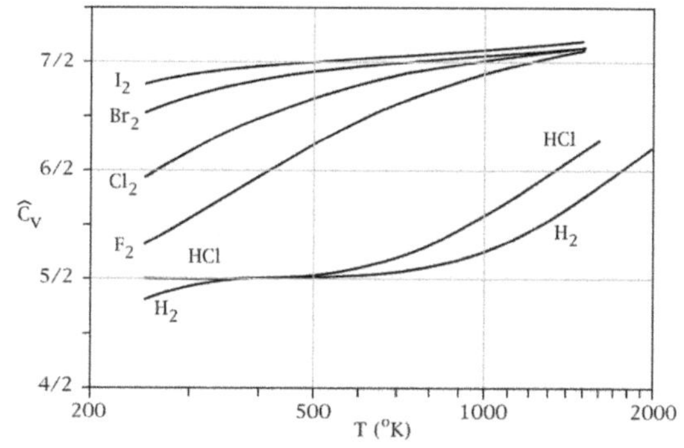

Constant volume specific heat capacity of diatomic gases (real gases) between about 200 K and 2000 K. This temperature range is not large enough to include both quantum transitions in all gases. Instead, at 200 K, all but hydrogen are fully rotationally excited, so all have at least 5/2 R heat capacity. (Hydrogen is already below 5/2, but it will require cryogenic conditions for even H2 to fall to 3/2 R). Further, only the heavier gases fully reach 7/2 R at the highest temperature, due to the relatively small vibrational energy spacing of these molecules. HCl and H_2 begin to make the transition above 500 K, but have not achieved it by 1000 K, since their vibrational energy level spacing is too wide to fully participate in heat capacity, even at this temperature.

General Gas Phase

The specific heat of the gas is best conceptualized in terms of the degrees of freedom of an individual molecule. The different degrees of freedom correspond to the different ways in which the molecule may store energy. The molecule may store energy in its translational motion according to the formula:

$$E = \frac{1}{2}m\left(v_x^2 + v_y^2 + v_z^2\right)$$

where m is the mass of the molecule and $[v_x, v_y, v_z]$ is velocity of the center of mass of the molecule. Each direction of motion constitutes a degree of freedom, thus there are three translational degrees of freedom.

In addition, a molecule may have rotational motion. The kinetic energy of rotational motion is generally expressed as

$$E = \frac{1}{2}\left(I_1\omega_1^2 + I_2\omega_2^2 + I_3\omega_3^2\right)$$

where I is the moment of inertia tensor of the molecule, and $[\omega_1, \omega_2, \omega_3]$ is the angular velocity pseudo-vector (in a coordinate system aligned with the principal axes of the molecule). In general, then, there will be three additional degrees of freedom corresponding to the rotational motion of the molecule, (For linear molecules one of the inertia tensor terms vanishes and there are only two rotational degrees of freedom). The degrees of freedom corresponding to translations and rotations are called the rigid degrees of freedom, since they do not involve any deformation of the molecule.

The motions of the atoms in a molecule which are not part of its gross translational motion or rotation may be classified as vibrational motions. It can be shown that if there are n atoms in the molecule, there will be as many as $v = 3n - 3 - n_r$ vibrational degrees of freedom, where n_r is the number of rotational degrees of freedom. A vibrational degree of freedom corresponds to a specific way in which all the atoms of a molecule can vibrate. The actual number of possible vibrations may be less than this maximal one, due to various symmetries.

For example, triatomic nitrous oxide N_2O will have only 2 degrees of rotational freedom (since it is a linear molecule) and contains n=3 atoms: thus the number of possible vibrational degrees of freedom will be v = (3·3) − 3 − 2 = 4. There are four ways or "modes" in which the three atoms can vibrate, corresponding to *1)* A mode in which an atom at each end of the molecule moves away from, or towards, the center atom at the same time, *2)* a mode in which either end atom moves asynchronously with regard to the other two, and *3)* and *4)* two modes in which the molecule bends out of line, from the center, in the two possible planar directions that are orthogonal to its axis. Each vibrational degree of freedom confers TWO total degrees of freedom, since vibrational energy mode partitions into 1 kinetic and 1 potential mode. This would give nitrous oxide 3 translational, 2 rotational, and 4 vibrational modes (but these last giving 8 vibrational degrees of freedom), for storing energy. This is a total of $f = 3 + 2 + 8 = 13$ total energy-storing degrees of freedom, for N_2O.

For a bent molecule like water H_2O, a similar calculation gives 9 − 3 − 3 = 3 modes of vibration, and 3 (translational) + 3 (rotational) + 6 (vibrational) = 12 degrees of freedom.

The Storage of Energy into Degrees of Freedom

If the molecule could be entirely described using classical mechanics, then the theorem of equipartition of energy could be used to predict that each degree of freedom would have an average energy in the amount of $(1/2)kT$ where k is Boltzmann's constant and T is the temperature. Our calculation of the constant-volume heat capacity would be straightforward. Each molecule would be holding, on average, an energy of $(f/2)kT$ where f is the total number of degrees of freedom in the molecule. Note that $Nk = R$ if N is Avogadro's number, which is the case in considering the heat capacity of a mole of molecules. Thus, the total internal energy of the gas would be $(f/2)NkT$ where N is the total number of molecules. The heat capacity (at constant volume) would then be a constant $(f/2)Nk$ the mole-specific heat capacity would be $(f/2)R$ the molecule-specific heat capacity would be $(f/2)k$ and the dimensionless heat capacity would be just $f/2$. Here again, each vibrational degree of freedom contributes 2f. Thus, a mole of nitrous oxide would have a total constant-volume heat capacity (including vibration) of (13/2) R by this calculation.

In summary, the molar heat capacity (mole-specific heat capacity) of an ideal gas with f degrees of freedom is given by

$$C_{V,m} = \frac{f}{2}R$$

This equation applies to all polyatomic gases, if the degrees of freedom are known.

The constant-pressure heat capacity for any gas would exceed this by an extra R. As example C_p would be a total of (15/2)R/mole for nitrous oxide.

The Effect of Quantum Energy Levels in Storing Energy in Degrees of Freedom

The various degrees of freedom cannot generally be considered to obey classical mechanics, however. Classically, the energy residing in each degree of freedom is assumed to be continuous—it can take on any positive value, depending on the temperature. In reality, the amount of energy that may reside in a particular degree of freedom is quantized: It may only be increased and decreased in finite amounts. A good estimate of the size of this minimum amount is the energy of the first excited state of that degree of freedom above its ground state. For example, the first vibrational state of the hydrogen chloride (HCl) molecule has an energy of about 5.74×10^{-20} joule. If this amount of energy were deposited in a classical degree of freedom, it would correspond to a temperature of about 4156 K.

If the temperature of the substance is so low that the equipartition energy of $(1/2)kT$ is much smaller than this excitation energy, then there will be little or no energy in this degree of freedom. This degree of freedom is then said to be "frozen out". As mentioned above, the temperature corresponding to the first excited vibrational state of HCl is about 4156 K. For temperatures well below this value, the vibrational degrees of freedom of the HCl molecule will be frozen out. They will contain little energy and will not contribute to the thermal energy or the heat capacity of HCl gas.

Energy Storage Mode "Freeze-out" Temperatures

It can be seen that for each degree of freedom there is a critical temperature at which the degree of freedom "unfreezes" and begins to accept energy in a classical way. In the case of translational degrees of freedom, this temperature is that temperature at which the thermal wavelength of the molecules is roughly equal to the size of the container. For a container of macroscopic size (e.g. 10 cm) this temperature is extremely small and has no significance, since the gas will certainly liquify or freeze before this low temperature is reached. For any real gas translational degrees of freedom may be considered to always be classical and contain an average energy of $(3/2)kT$ per molecule.

The rotational degrees of freedom are the next to "unfreeze". In a diatomic gas, for example, the critical temperature for this transition is usually a few tens of kelvins, although with a very light molecule such as hydrogen the rotational energy levels will be spaced so widely that rotational heat capacity may not completely "unfreeze" until considerably higher temperatures are reached. Finally, the vibrational degrees of freedom are generally the last to unfreeze. As an example, for diatomic gases, the critical temperature for the vibrational motion is usually a few thousands of kelvins, and thus for the nitrogen in our example at room temperature, no vibration modes would be excited, and the constant-volume heat capacity at room temperature is $(5/2)R$/mole, not $(7/2)R$/mole. As seen above, with some unusually heavy gases such as iodine gas I_2, or bromine gas Br_2, some vibrational heat capacity may be observed even at room temperatures.

It should be noted that it has been assumed that atoms have no rotational or internal degrees of freedom. This is in fact untrue. For example, atomic electrons can exist in excited states and even the atomic nucleus can have excited states as well. Each of these internal degrees of freedom are assumed to be frozen out due to their relatively high excitation energy. Nevertheless, for sufficiently high temperatures, these degrees of freedom cannot be ignored. In a few exceptional cases, such molecular electronic transitions are of sufficiently low energy that they contribute to heat capacity at room temperature, or even at cryogenic temperatures. One example of an electronic transition degree of freedom which contributes heat capacity at standard temperature is that of nitric oxide (NO), in which the single electron in an anti-bonding molecular orbital has energy transitions which contribute to the heat capacity of the gas even at room temperature.

An example of a nuclear magnetic transition degree of freedom which is of importance to heat capacity, is the transition which converts the spin isomers of hydrogen gas (H_2) into each other. At room temperature, the proton spins of hydrogen gas are aligned 75% of the time, resulting in *orthohydrogen* when they are. Thus, some thermal energy has been stored in the degree of freedom available when *parahydrogen* (in which spins are anti-aligned) absorbs energy, and is converted to the higher energy ortho form. However, at the temperature of liquid hydrogen, not enough heat energy is available to produce orthohydrogen (that is, the transition energy between forms is large enough to "freeze out" at this low temperature), and thus the parahydrogen form predominates. The heat capacity of the transition is sufficient to release enough heat, as orthohydrogen converts to the lower-energy parahydrogen, to boil the hydrogen liquid to gas again, if this evolved heat is not removed with a catalyst after the gas has been cooled and condensed. This example also illustrates the fact that some modes of storage of heat may not be in constant equilibrium with each other in substances, and heat absorbed or released from such phase changes may "catch up" with temperature changes of substances, only after a certain time. In other words, the heat evolved and absorbed from the ortho-para isomeric transition contributes to the heat capacity of hydrogen on long time-scales, but not on *short* time-scales. These time scales may also depend on the presence of a catalyst.

Less exotic phase-changes may contribute to the heat-capacity of substances and systems, as well, as (for example) when water is converted back and forth from solid to liquid or gas form. Phase changes store heat energy entirely in breaking the bonds of the potential energy interactions between molecules of a substance. As in the case of hydrogen, it is also possible for phase changes to be hindered as the temperature drops, thus they do not catch up and become apparent, without a catalyst. For example, it is possible to supercool liquid water to below the freezing point, and not observe the heat evolved when the water changes to ice, so long as the water remains liquid. This heat appears instantly when the water freezes.

Solid Phase

For matter in a crystalline solid phase, the Dulong–Petit law, which was discovered empirically, states that the molar heat capacity assumes the value 3 R. Indeed, for solid metallic chemical elements at room temperature, molar heat capacities range from about 2.8 R to 3.4 R. Large exceptions at the lower end involve solids composed of relatively low-mass, tightly bonded atoms, such as beryllium at 2.0 R, and diamond at only

0.735 R. The latter conditions create larger quantum vibrational energy spacing, thus many vibrational modes have energies too high to be populated (and thus are "frozen out") at room temperature. At the higher end of possible heat capacities, heat capacity may exceed R by modest amounts, due to contributions from anharmonic vibrations in solids, and sometimes a modest contribution from conduction electrons in metals. These are not degrees of freedom treated in the Einstein or Debye theories.

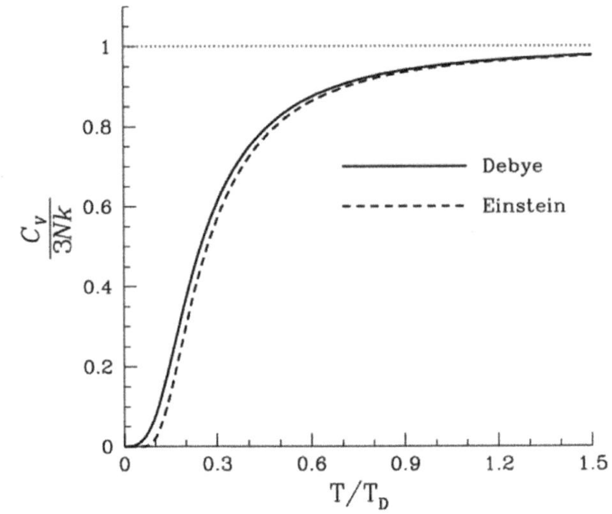

The dimensionless heat capacity divided by three, as a function of temperature as predicted by the Debye model and by Einstein's earlier model. The horizontal axis is the temperature divided by the Debye temperature. Note that, as expected, the dimensionless heat capacity is zero at absolute zero, and rises to a value of three as the temperature becomes much larger than the Debye temperature. The red line corresponds to the classical limit of the Dulong–Petit law

The theoretical maximum heat capacity for multi-atomic gases at higher temperatures, as the molecules become larger, also approaches the Dulong–Petit limit of 3 R, so long as this is calculated per mole of atoms, not molecules. The reason for this behavior is that, in theory, gases with very large molecules have almost the same high-temperature heat capacity as solids, lacking only the (small) heat capacity contribution that comes from potential energy that cannot be stored between separate molecules in a gas.

The Dulong–Petit limit results from the equipartition theorem, and as such is only valid in the classical limit of a microstate continuum, which is a high temperature limit. For light and non-metallic elements, as well as most of the common molecular solids based on carbon compounds at standard ambient temperature, quantum effects may also play an important role, as they do in multi-atomic gases. These effects usually combine to give heat capacities lower than 3 R per mole of *atoms* in the solid, although in molecular solids, heat capacities calculated *per mole of molecules* in molecular solids may be more than 3 R. For example, the heat capacity of water ice at the melting point is about 4.6 R per mole of molecules, but only 1.5 R per mole of atoms. As noted, heat capacity values far lower than 3 R "per atom" (as is the case with diamond and beryllium) result from "freezing out" of possible vibration modes for light atoms at suitably low temperatures, just as happens in many low-mass-atom gases at room temperatures (where vibrational modes are all frozen out). Because of high crystal binding energies, the effects of vibrational mode freezing are observed in solids more often than liquids: for example the heat capacity of liquid water is twice that of ice at near the same temperature, and is again close to the 3 *R* per mole of atoms of the Dulong–Petit theoretical maximum.

Liquid Phase

A general theory of the heat capacity of liquids has not yet been achieved, and is still an active area of research. It was long thought that phonon theory is not able to explain the heat capacity of liquids, since liquids only sustain longitudinal, but not transverse phonons, which in solids are responsible for 2/3 of the heat capacity. However, Brillouin scattering experiments with neutrons and with X-rays, confirming an intuition of Yakov Frenkel, have shown that transverse phonons do exist in liquids, albeit restricted to frequencies above a threshold called the Frenkel frequency. Since most energy is contained in these high-frequency modes, a simple modification of the Debye model is sufficient to yield a good approximation to experimental heat capacities of simple liquids.

Amorphous materials can be considered a type of liquid. The specific heat of amorphous materials has characteristic discontinuities at the glass transition temperature. These discontinuities are frequently used to detect the glass transition temperature where a supercooled liquid transforms to a glass.

Table of Specific Heat Capacities

Note that the especially high molar values, as for paraffin, gasoline, water and ammonia, result from calculating specific heats in terms of moles of *molecules*. If specific heat is expressed per mole of *atoms* for these substances, none of the constant-volume values exceed, to any large extent, the theoretical Dulong–Petit limit of 25 $J \cdot mol^{-1} \cdot K^{-1}$ = 3 R per mole of atoms. Paraffin, for example, has very large molecules and thus a high heat capacity per mole, but as a substance it does not have remarkable heat capacity in terms of volume, mass, or atom-mol (which is just 1.41 R per mole of atoms, or less than half of most solids, in terms of heat capacity per atom).

In the last column, major departures of solids at standard temperatures from the Dulong–Petit law value of 3 R, are usually due to low atomic weight plus high bond strength (as in diamond) causing some vibration modes to have too much energy to be available to store thermal energy at the measured temperature. For gases, departure from 3 R per mole of atoms in this table is generally due to two factors: (1) failure of the higher quantum-energy-spaced vibration modes in gas molecules to be excited at room temperature, and (2) loss of potential energy degree of freedom for small gas molecules, simply because most of their atoms are not bonded maximally in space to other atoms, as happens in many solids.

Table of specific heat capacities at 25 °C (298 K) unless otherwise noted.

Notable minima and maxima are shown in maroon

Substance	Phase	Isobaric mass heat capacity c_P J·g⁻¹·K⁻¹	Isobaric molar heat capacity $C_{P,m}$ J·mol⁻¹·K⁻¹	Isochore molar heat capacity $C_{V,m}$ J·mol⁻¹·K⁻¹	Isobaric volumetric heat capacity $C_{P,v}$ J·cm⁻³·K⁻¹	Isochore atom-molar heat capacity in units of R $C_{V,am}$ atom-mol⁻¹
Air (Sea level, dry, 0 °C (273.15 K))	gas	1.0035	29.07	20.7643	0.001297	~ 1.25 R
Air (typical room conditions[A])	gas	1.012	29.19	20.85	0.00121	~ 1.25 R
Aluminium	solid	0.897	24.2		2.422	2.91 R
Ammonia	liquid	4.700	**80.08**		3.263	3.21 R
Animal tissue (incl. human)	mixed	3.5			3.7*	
Antimony	solid	0.207	25.2		1.386	3.03 R
Argon	gas	0.5203	20.7862	12.4717		1.50 R
Arsenic	solid	0.328	24.6		1.878	2.96 R
Beryllium	solid	1.82	16.4		3.367	1.97 R
Bismuth	solid	0.123	25.7		1.20	3.09 R
Cadmium	solid	0.231	26.02			3.13 R
Carbon dioxide CO_2	gas	0.839*	36.94	28.46		1.14 R
Chromium	solid	0.449	23.35			2.81 R
Copper	solid	0.385	24.47		3.45	2.94 R
Diamond	solid	0.5091	**6.115**		1.782	0.74 R
Ethanol	liquid	2.44	112		1.925	1.50 R
Gasoline (octane)	liquid	2.22	228		1.64	1.05 R
Glass	solid	0.84			2.1	
Gold	solid	0.129	25.42		2.492	3.05 R
Granite	solid	0.790			2.17	
Graphite	solid	0.710	8.53		1.534	1.03 R
Helium	gas	5.1932	20.7862	12.4717		1.50 R
Hydrogen	gas	**14.30**	28.82			1.23 R
Hydrogen sulfide H_2S	gas	1.015*	34.60			1.05 R
Iron	solid	0.412	25.09		3.537	3.02 R
Lead	solid	0.129	26.4		1.44	3.18 R

Lithium	solid	3.58	24.8		1.912	2.98 R
Lithium at 181 °C	liquid	**4.379**	30.33		2.242	3.65 R
Magnesium	solid	1.02	24.9		1.773	2.99 R
Mercury	liquid	0.1395	27.98		1.888	3.36 R
Methane at 2 °C	gas	2.191	35.69			0.85 R
Methanol	liquid	2.14	68.62			1.38 R
Molten salt (142–540 °C)	liquid	1.56			2.62	
Nitrogen	gas	1.040	29.12	20.8		1.25 R
Neon	gas	1.0301	20.7862	12.4717		1.50 R
Oxygen	gas	0.918	29.38	21.0		1.26 R
Paraffin wax $C_{25}H_{52}$	solid	2.5 (ave)	**900**		2.325	1.41 R
Polyethylene (rotomolding grade)	solid	2.3027				
Silica (fused)	solid	0.703	42.2		1.547	1.69 R
Silver	solid	0.233	24.9		2.44	2.99 R
Sodium	solid	1.230	28.23			3.39 R
Steel	solid	0.466			3.756	
Tin	solid	0.227	27.112		1.659	3.26 R
Titanium	solid	0.523	26.060		2.6384	3.13 R
Tungsten	solid	0.134	24.8		2.58	2.98 R
Uranium	solid	**0.116**	27.7		2.216	3.33 R
Water at 100 °C (steam)	gas	**2.080**	**37.47**	**28.03**		1.12 R
Water at 25 °C	liquid	**4.1813**	**75.327**	**74.53**	**4.1796**	3.02 R
Water at 100 °C	liquid	**4.1813**	**75.327**	**74.53**	**4.2160**	3.02 R
Water at −10 °C (ice)	solid	**2.05**	**38.09**		**1.938**	1.53 R
Zinc	solid	0.387	25.2		2.76	3.03 R

A Assuming an altitude of 194 metres above mean sea level (the world–wide median altitude of human habitation), an indoor temperature of 23 °C, a dewpoint of 9 °C (40.85% relative humidity), and 760 mm–Hg sea level–corrected barometric pressure (molar water vapor content = 1.16%).

*Derived data by calculation. This is for water-rich tissues such as brain. The whole-body average figure for mammals is approximately 2.9 $J \cdot cm^{-3} \cdot K^{-1}$

Mass Heat Capacity of Building Materials

(Usually of interest to builders and solar designers)

Mass heat capacity of building materials		
Substance	**Phase**	c_P $J \cdot g^{-1} \cdot K^{-1}$
Asphalt	solid	0.920
Brick	solid	0.840
Concrete	solid	0.880
Glass, silica	solid	0.840
Glass, crown	solid	0.670
Glass, flint	solid	0.503
Glass, pyrex	solid	0.753
Granite	solid	0.790
Gypsum	solid	1.090
Marble, mica	solid	0.880
Sand	solid	0.835
Soil	solid	0.800
Water	liquid	4.1813
Wood	solid	1.7 (1.2 to 2.9)

Thermal Conductivity

- If there is a temperature gradient, heat will flow from higher to lower temperature region. This is Thermal conduction.
- The ability of a material to transfer the heat is the Thermal conductivity, *k*.

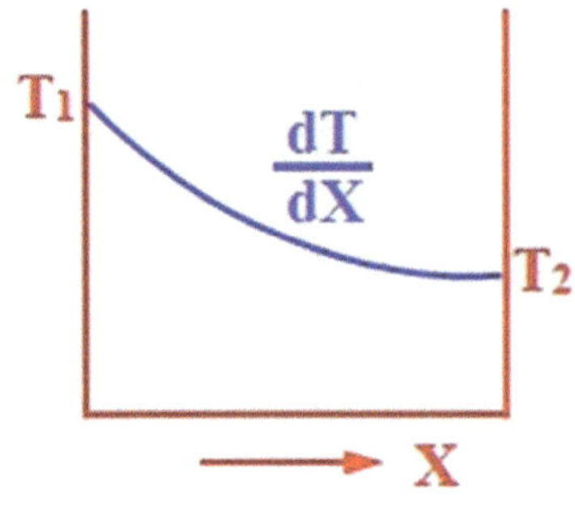

$$q = -k\frac{dT}{dX}$$

q is steady state heat flux i.e. heat flow per unit area per unittime (W/m^2)

k is thermal conductivity of the medium (W/m-K)

dT/dX is the thermal gradient in the medium.

Conduction Mechanism

- Atoms vibrate about their equilibrium positions with high frequency and low amplitudes. Amplitude increases with rise in temperature.

- The vibrations of adjacent atoms are coupled due to atomic bonding and this leads to generation of elastic waves which move through the lattice at the velocity of sound and thus carries the heat.
- Each quantum of the wave is known as phonon.

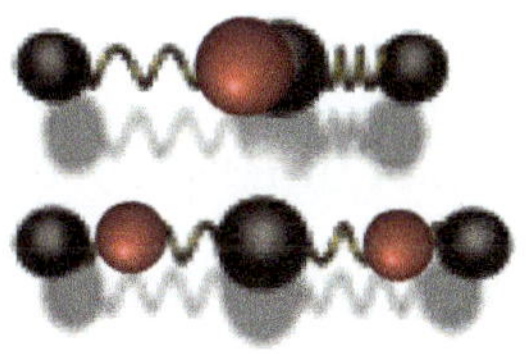

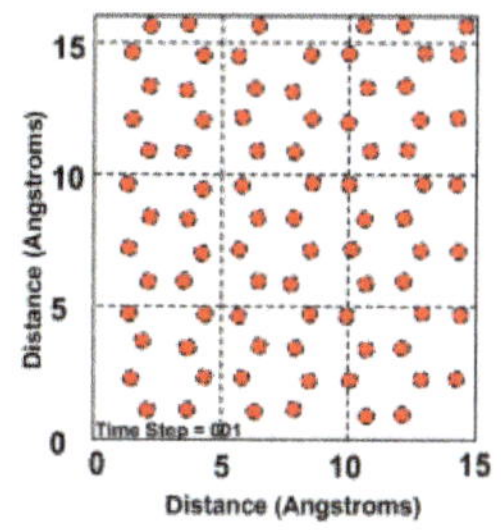

- Free electrons gain kinetic energy in the hotter region and move towards the colder region thus transferring the heat.
- Therefore, thermal conductivity $k = k_l$ (lattice) + k_e (electron)
- Since a large number free valence electrons are available in metals, the electron mechanism is much more efficient. This imparts great thermal conductivity that metals are known for.
- Thermal and electrical conductivities in metals are related by *Wiedemann–Franz law:* $k / \sigma T$, L is a constant, σ is electrical conductivity.
- Ceramics do not have free electrons as all electrons are tightly bound in the atomic bonds and hence, are poor conductors of heat.
- Polymers conduct heat by vibrational and rotational motion of chain molecules and hence, are poor conductors of heat.

Thermal Conductivity

Thermal conductivity of ceramics generally decreases with increasing temperature due to phonon scattering. At very high temperature it increases again due to change in heat transfer mode from conduction to radiation.

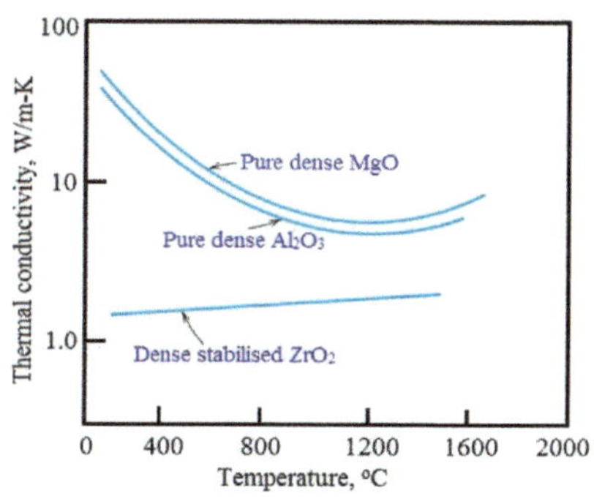

Determination of Thermal Conductivity

Experimental set up

- Heat generation/conduction unit with controls
- Thermocouples and Data logger
- Maintain two ends of the sample at constant temperatures
- Insert five to six thermocouples along the length at different locations
- Plot temperature as a function of distance and find dT/dX from the slope.
- Find the thermal conductivity, k (q = - k dT/dX) [Experimental techniques to find q can be found at

Thermal Expansion

- Most of the solids expand when heated. This is known as thermal expansion.
- It can be expressed as $\frac{\Delta L}{L_0} = \alpha_l \Delta T$
  Where, ΔL is the change in length due to a temperature rise of ΔT . α_l is known as linear coefficient of thermal expansion (CTE).
- α_1 (°c^{-1}) is a material property which depends on the type of atomic bonding. The extent to which a material expands on heating will depend on its α_1 .
- The atomic mechanism of thermal expansion can be viewed as an increase in the inter-atomic separation.
- Therefore, it will depend on the shape of the energy vs. inter- atomic distance curve.

- The energy and vibrational amplitude (width of the potential energy trough) increase with increasing temperature and so does the interatomic separation (indicated by the open circles).
- For a material with a broader potential curve, the increase in the interatomic distance is more (Fig. a) and hence thermal expansion is more.
- The increase in atomic distance and hence, the expansion is much lower for a deep and narrow potential trough (Ceramics).

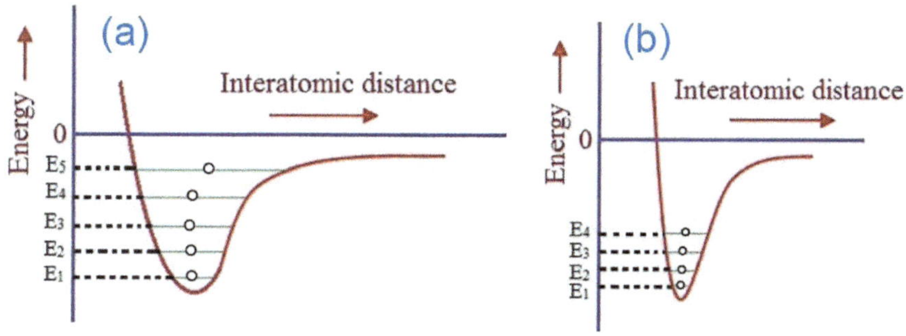

Low/Zero Thermal Expansion

- Coefficient of thermal expansion (CTE) for metals is in the range of $5 \times 10^{-6} - 25 \times 10^{-6}$ /°C.
- For a typical ceramic like Al_2O_3 CTE = 7.6×10^{-6} /°C
- There is a class of materials which have very low or near-zero thermal expansion.
- Invar (64Fe – 36Ni) has CTE of 1.6×10^{-6} /°C (Up to 230 °C, the temperature can be increased by heat treatment).
- Super Invar (63Fe-32Ni-5Co) – 0.72×110^{-6} /°C.
- This is believed to be caused by magnetostriction a phenomena which lead to volume change on magnetization.

Negative Thermal Expansion

- Some materials contract on heating (negative CTE). Zirconium tungstate (ZrW_2O_8) for example contracts continuously from 2 to 1050 K.
- A composite (mix) of positive and negative expansion materials may give rise to a zero expansion material.

Thermal Properties of Some Materials

Material	c_p (J/kg-K)	α_l (°C^{-1} x 10^{-6})	k (W/m-K)
Metals			
Alumnium	900	23.6	247
Copper	386	17.0	398
Silver	235	19.7	428
Steel	502	16	15.9
Super Invar	500	0.72	10

Ceramics			
Alumina (Al_2O_3)	775	7.6	39
Fused Silica (SiO_2)	740	0.4	1.4
Pyrex glass	850	3.3	1.4
Polymers			
Polyethylene	1850	106 - 198	0.50
Polystyrene	1170	90-150	0.13
Teflon	1050	126-216	0.25

Thermal Stress

Thermal stresses arise due to –

- Constrained expansion or contraction e.g. heating or cooling a rod with fixed rigid ends.
- Uneven heating/cooling
- Thermal expansion mismatch inside the solid.

Thermal stress σ due to temperature change from T_0 to T_f

$$\sigma = E\alpha_1(T_o - T_f) = E\alpha_1\Delta T$$

E is the elastic modulus.

Upon heating, the stress is compressive and tensile while cooling if the expansion/ contraction is restrained.

Example

A steel rod is to be used with its ends held rigid. What is the maximum temperature the rod can be heated to without the compressive stress in it exceeding 180 MPa. Elastic modulus of the rod E = 190 GPa.

Solution:

$$\sigma \le -180\times10^6\, Pa(Compressive) = E\alpha_l(T_o - T_f)$$

$$\alpha_l \text{ for steel} \cong 14\times10^{-6}\, {}^oC^{-1}. T_o = \text{room temperature} = 25\,^0C$$

$$T_f = 25 + \frac{180\times10^6}{190\times10^9\times14\times10^{-6}} = 92^o C$$

Thermal Shock

Thermal stresses might cause fracture in brittle materials like ceramics due to rapid heating or cooling if the expansion/contraction is restrained. This is known as thermal shock.

The ability of material to withstand such shocks is known as thermal shock resistance (TSR)

$$TSR \cong \frac{\sigma_f k}{E\alpha_l},\ \sigma_f \textit{ is the fracture stress.}$$

Thermal shock can be prevented by controlling the external conditions like lowering heating and cooling rates and controlling the thermal/mechanical parameters such as CTE and fracture stress as per the equation above.

Electrical Conduction

Ohm's law $V = IR$

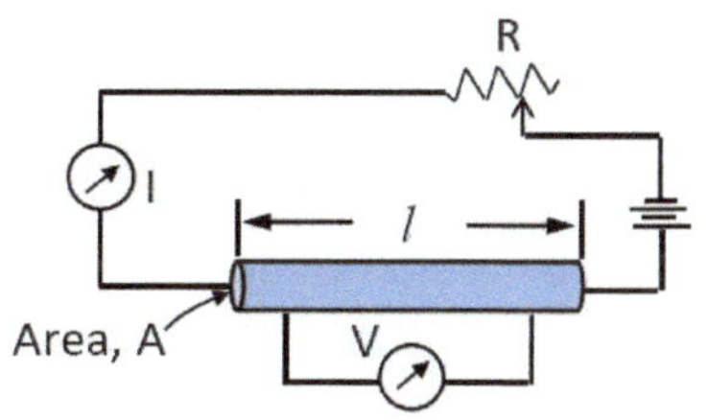

where I is current (Ampere), V is voltage (Volts) and R is the resistance (Ohms or Ω) of the conductor

Resistivity

Resistivity, ρ = RA/l (Ω -m), where A is the area and l is the length of the conductor.

Electrical Conductivity

Conductivity, $\sigma = 1/\rho = l/RA$ (Ω -m)-1

Band Theory

- Electrons occupy energy states in atomic orbitals
- When several atoms are brought close to each other in a solid these energy states split in to a series of energy states (molecular orbitals).
- The spacing between these states are so small that they overlap to form an en-

ergy band.

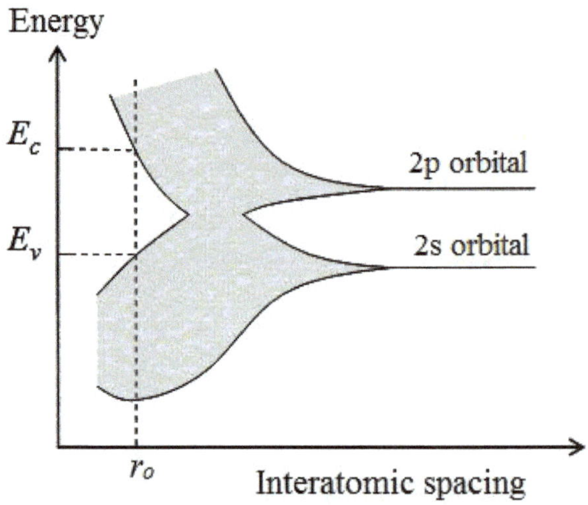

- The furthest band from the nucleus is filled with valence electrons and is called the valence band.
- The empty band is called the conduction band.
- The energy of the highest filled state is called Fermi energy.
- There is a certain energy gap, called band gap, between valence and conduction bands.
- Primarily four types of band structure exist in solids.

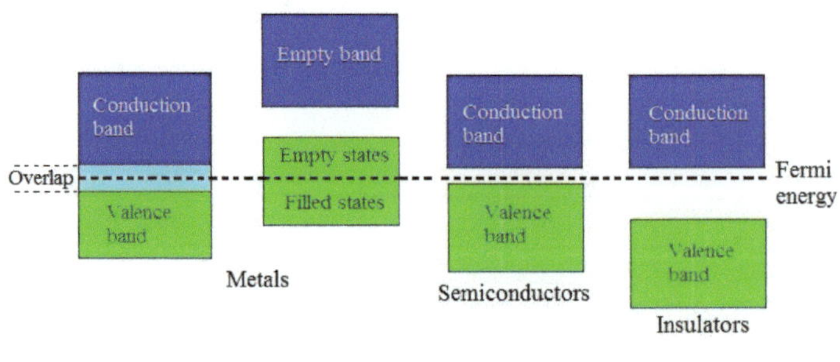

- In metals the valence band is either partially filled (Cu) or the valence and conduction bands overlap (Mg).
- Insulators and semiconductors have completely filled valence band and empty conduction band.
- It is the magnitude of band gap which separates metals, semiconductors and insulators in terms of their electrical conductivity.
- The band gap is relatively smaller in semiconductors while it is very large in insulators.

Conduction Mechanism

- An electron has to be excited from the filled to the empty states above Fermi level (E_f) for it to become free and a charge carrier.
- In metals large number of free valence electrons are available and they can be easily excited to the empty states due to their band structure
- On the other hand a large excitation energy is needed to excite electrons in Insulators and semiconductors due the large band gap.

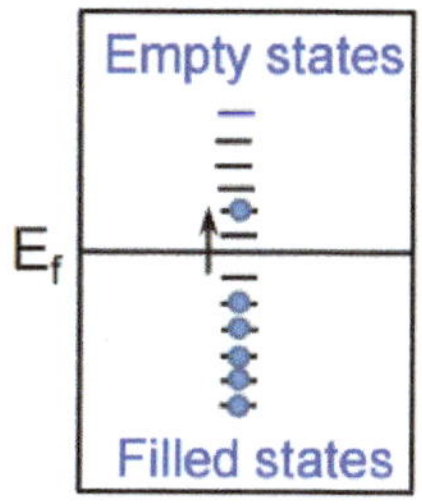

Conduction in Metals

Intrinsic Semiconductors

- Semiconductors like Si and Ge have relatively narrow band gap generally below 2 eV.
- Therefore, it is possible to excite electrons from the valence to the conduction band. This is called intrinsic semi conductivity.
- Every electron that is excited to the conduction band leaves behind a hole in the valence band.
- An electron can move in to a *hole* under an electrical potential and thus holes are also charge carriers.

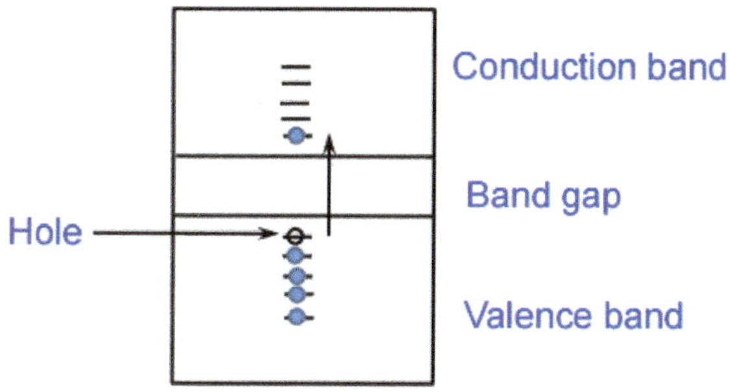

Intrinsic Conductivity

- Electrical conductivity of a conductor primarily depends on two parameters – charge carrier concentration, *n*, and carrier mobility, μ . Conductivity, $\sigma = n|e|\mu$ $|e|$ *is absolute charge* $(1.6\times10^{-19}C)$.

- Intrinsic semiconductors have two types charge carriers, namely electrons and holes

$$\sigma = n|e|\mu_e + p|e|\mu_\eta$$

where, n and p are concentration of electron and hole charge carriers respectively and μ_e and μ_η are their mobility.

- Since each electron excited to conduction band leaves behind a hole in the valence band, $n = p = n_i$ and $\sigma = n|e|(\mu_e + \mu_\eta) = p|e|(\mu_e + \mu_\eta) = n_i|e|(\mu_e + \mu_\eta)$

Extrinsic Semiconductors

- The conductivity is enhanced by adding impurity atoms (dopant) in extrinsic semi conductors. All semi conductors for practical purposes are extrinsic.
- A higher valence dopant e.g. P (5+) in Si (4+) creates an extra electron (n-type) while a lower valence dopant like B (3+) creates a hole (p-type) as shown in the atomic bonding model below.
- This increases the charge carrier concentration and hence the enhancement in conductivity.

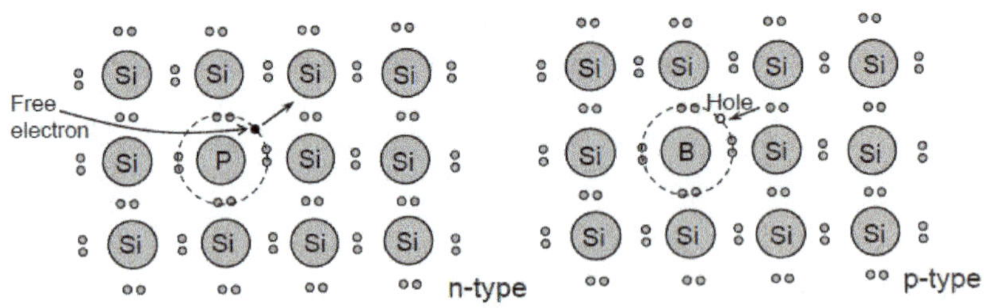

- The band theory model of n-type and p-type extrinsic semiconductors are shown below.
- In n-type, for each impurity atom one energy state (known as Donor state) is introduced in the band gap just below the conduction band.
- In p-type, for each impurity atom one energy state (known as acceptor state) is introduced in the band gap just above the valence band.

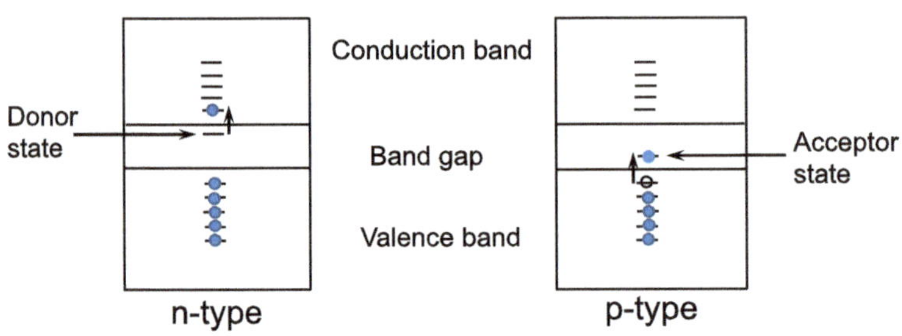

Extrinsic Conductivity

- Large number of electrons can be excited from the donor state by thermal energy in n-type extrinsic semiconductors.
- Hence, number of electrons in the conduction band is far greater than number of holes in the valence band, i.e. $n >> p$ and

$$\sigma = n|e|\mu_e$$

- In p-type conductors, on the other hand, number of holes is much greater than electrons ($p >> n$) due to the presence of the acceptor states.

$$\sigma = p|e|\mu_h$$

Effect of Temperature

Metals

Increasing temperature causes greater electron scattering due to increased thermal vibrations of atoms and hence, resistivity, ρ, (reciprocal of conductivity) of metals increases (conductivity decreases) linearly with temperature.

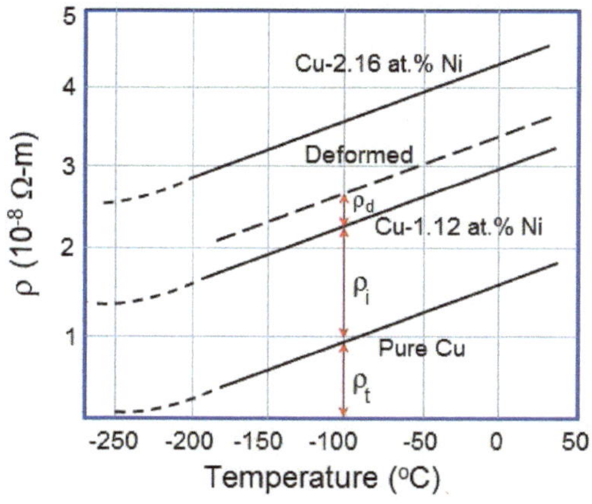

- The resistivity of metals depends on two other factors namely, impurity level and plastic deformation as these generate scattering centers for electrons.
- Increase in impurity level results in more scattering centers and decreases the conductivity.
- Similarly plastic deformation introduces more dislocations which act as scattering centers and increase the resistivity.

$$\rho_{total} = \rho_t + \rho_i + \rho_d$$

Intrinsic Semiconductors

- In intrinsic semiconductors the carrier concentration increases with temperature as more and more electrons are excited due to the thermal energy.

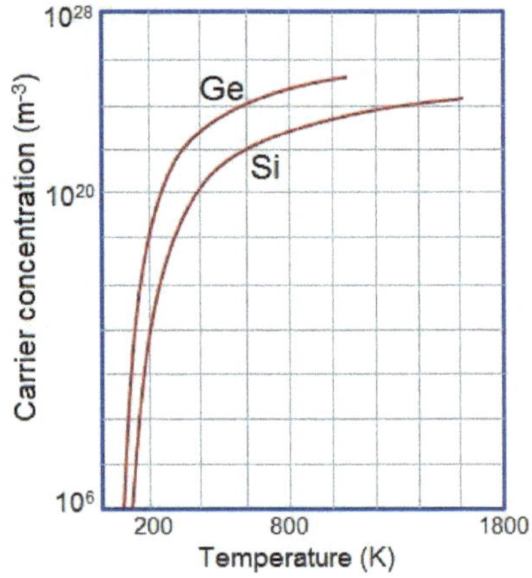

Extrinsic Semiconductors

- Temperature dependence of extrinsic semiconductors, on the other hand is totally different.
- For example, an n-type conductor exhibits three regions in the temperature vs. carrier concentration curve.

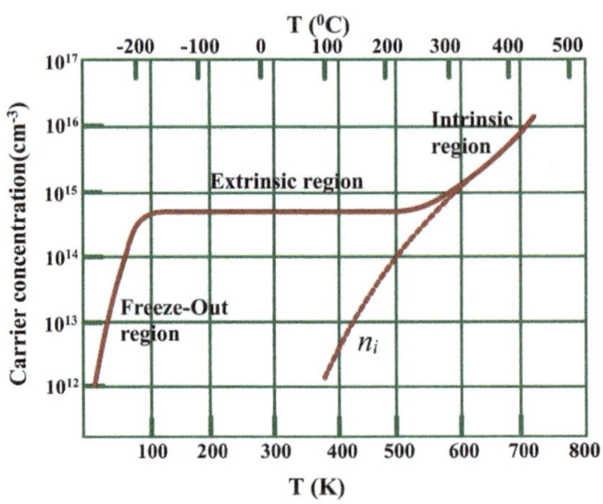

- In the low temperature region known as *Freeze-out* region, the charge carriers cannot be excited from the donor level to conduction band due to insufficient thermal energy.
- In the intermediate temperature range (150–450K) almost all the donor atoms are ionized and electron concentration is approximately equal to donor content. This region is known as *Extrinsic region.*

- In the high temperature region sufficient thermal energy is available for electrons to get excited from the valence to the conduction band and hence it behaves like an intrinsic semi conductor.

Electrical Properties of Some Metals at RT

Metal	Conductivity (Ω^{-1}-m^{-1})	Resistivity (Ω-m)
Silver	6.8×10^{7}	1.59×10^{-8}
Copper	6.0×10^{7}	1.68×10^{-8}
Gold	4.3×10^{7}	2.44×10^{-8}
Aluminum	3.8×10^{7}	2.82×10^{-8}
Nickel	1.43×10^{7}	6.99×10^{-8}
Iron	1.0×10^{7}	9.0×10^{-8}
Platinum	0.94×10^{7}	1.06×10^{-7}

Electrical Properties of Some Semi Conductors

Material	Band gap (eV)	Conductivity (Ω^{-1}-m^{-1})	μ_n (m^2/V-s)	μ_p (m^2/V-s)
Si	1.11	4×10^{-4}	0.14	0.05
Ge	0.67	2.2	0.38	0.18
GaP	2.25	-	0.03	0.015
GaAs	1.42	1×10^{-6}	0.85	0.04
InSb	0.17	2×10^{4}	7.7	0.07
CdS	2.40	-	0.03	-
ZnTe	2.26	-	0.03	0.01

Dielectric Property

- A dielectric material is an insulating material which can separate positive and negatively charged entities.
- Dielectric materials are used in capacitors to store the electrical energy.

Capacitance

- Capacitance, C, is related to charge stored, Q, between two oppositely charged layers subjected to a voltage V. $C = Q/V$
- If two parallel plates of area, A, are separated by a distance l in vacuum, then $C =_{\varepsilon o} A / I$. ε_o, permittivity of vacuum = 8.85×10^{-12} F/m.

- If a dielectric material is present between the plates, $C = \varepsilon A / I$, ε is the permittivity of the dielectric medium.
- Relative permittivity $\varepsilon_r = \varepsilon / \varepsilon_o$, also known as dielectric constant.

Capacitance and Polarization

- The orientation of a dipole along the applied electric field is called polarization (*P*).
- It causes charge density to increase over that of a vacuum due to the presence of the dielectric material so that $D = \varepsilon_o \xi + P$. ξ is the electric field.
- *D* is surface charge density of a capacitor, also called dielectric displacement.

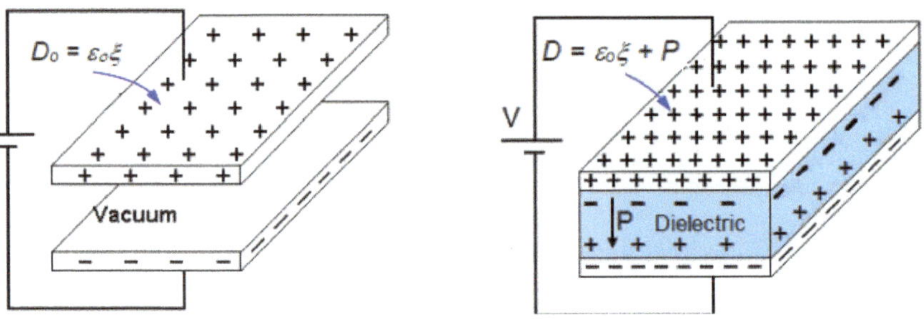

Types of Polarization

- Four types of polarization: Electronic, Ionic, Orientation, and Space charge (interfacial).
- Electronic polarization is due to displacement of the centre of the electron cloud around the nucleus under the applied field.
- Ionic polarization occurs in ionic material as the applied electric filed displaces the cations and anions in opposite directions resulting in a net dipole moment.
- Orientation polarization can only occur in materials having permanent dipole moments. The rotation of the permanent moment in the direction of the applied field causes the polarization in this case.
- Space charges polarization arises from accumulation of charge at interfaces in a heterogeneous material consisting of more than one phase having different resistivity.

Ferro-electricity

- Ferro-electricity is defined as the spontaneous alignment of electric dipoles in the absence of an external field.
- The spontaneous polarization results from relative displacement of cations

and anions from their symmetrical positions. Therefore, ferroelectric materials must posses permanent dipoles.

- Examples of ferroelectric materials: $BaTiO_3$, Rochelle salt ($NaKC_4H_4O_6.4H_2O$), potassium dihydrogen phosphate (KH_2PO_4), potassium niobate ($KNbO_3$), lead zirconate titanate [Pb (ZrO_3, TiO_3)].
- These materials have extremely high dielectric constants at relatively low applied field frequencies. Hence, capacitors made from ferroelectric materials are smaller than those made from other dielectric materials.

Piezoelectricity

- Piezo-electricity is defined as conversion of electrical energy into mechanical strain and vice versa.
- It arises due to polarization induced by an external force. Thus by reversing the direction of external force, direction of the established field can be reversed i.e. the application of an external electric field alters the net dipole length causing a dimensional change.
- Application for these materials includes microphones, ultrasonic generators, sonar detectors, and mechanical strain gauges.
- Examples: Barium titanate, lead titanate, lead zirconate ($PbZrO_3$),ammonium dihydrogen phosphate ($NH_4H_2PO_4$), and quartz.

Magnetic Properties

- Magnetic field is a force which is generated due to energy change in a volume of space.
- A magnetic field is produced by an electrical charge in motion e.g. current flowing in a conductor, orbital movement and spin of electrons.
- The magnetic field can be described by imaginary lines as shown in the figure below for a magnet and a current loop.

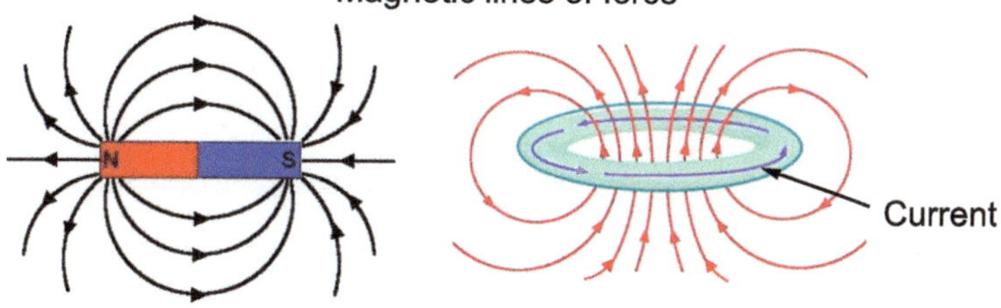

Magnetic Field Strength

- If a magnetic field, H, is generated by a cylindrical coil (solenoid) of n turns and length l, $H = nI/l$ (A/m)
- Magnetic flux density, B: It is the magnitude of the field strength within a substance subjected to a field H

  $B = \mu H$ (*Tesla or Weber / m^2*)

- μ, called the permeability, is the measure of the degree to which a material can be magnetized.
- In vacuum $B = \mu_0 H$. μ_0 is the permeability of vacuum and is a universal constant. $\mu_0 = 4\pi \times 10^{-7}$(H/m). $\mu_r = \mu / \mu_0$ is the relative permeability.

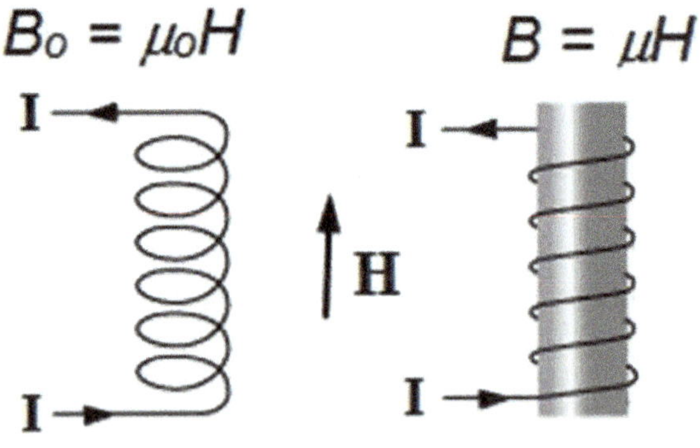

Magnetic Moments

- Being a moving charge, electrons produce a small magnetic field having a magnetic moment along the axis of rotation.
- The spin of electrons also produces a magnetic moment along the spin axis.
- Magnetism in a material arises due to alignment of magnetic moments.

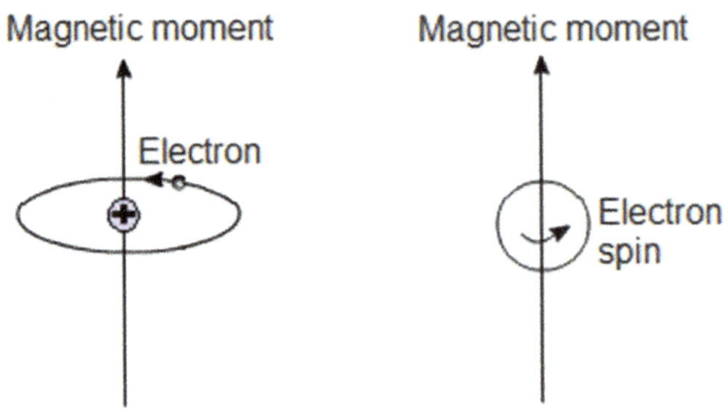

Magnetic Dipole and Monopole

- Analogous to electric dipole, a magnetic dipole can be defined as two monopoles of opposite and equal strength separated by a certain distance.
- A magnetic monopole, however, is not observed in nature.
- If there are *N* monopoles each located at a point given by a vector $\bar{a}$, then the magnetic dipole moment can be defined a vector, $\bar{u}$

$$\bar{u} = \sum_{i=1}^{N} \bar{m}_i \bar{a}_i$$

- Two monopoles of strength +m and −m separated by distance *l*, will give a dipole $\bar{u} = m\bar{a}_1 - m\bar{a}_2 = \text{m}(\bar{a}_1 - \bar{a}_2) = m\bar{l}$
- Abar magnet can be thought of consisting of two opposite and equal poles at its two ends.

Magnetization

- With the application of a magnetic field magnetic moments in a material tend to align and thus increase the magnitude of the field strength.
- This increase is given by the parameter called magnetization, M, such that $B = \mu_0 H + \mu_0 M$.

$$M = \chi_m H$$

χ_m is called magnetic susceptibility

$$\chi_m = \mu_r - 1$$

Magnetism

- Depending on the existence and alignment of magnetic moments with or without application of magnetic field, three types of magnetism can be defined.

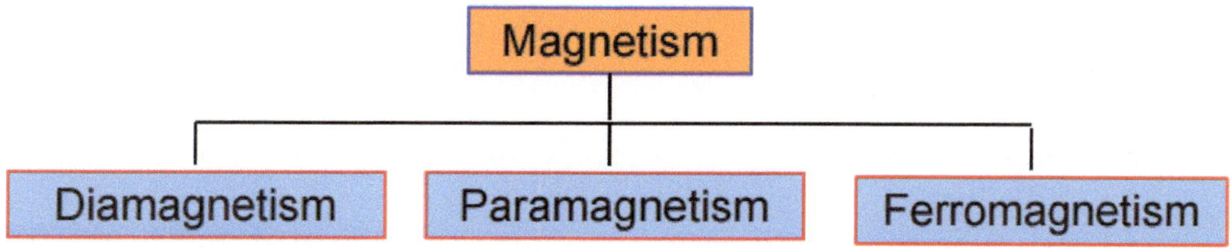

Diamagnetism

Pyrolytic carbon has one of the largest diamagnetic constants of any room temperature material. Here a pyrolytic carbon sheet is levitated by its repulsion from the strong magnetic field of neodymium magnets.

Diamagnetic materials are repelled by a magnetic field; an applied magnetic field creates an induced magnetic field in them in the opposite direction, causing a repulsive force. In contrast, paramagnetic and ferromagnetic materials are attracted by a magnetic field. Diamagnetism is a quantum mechanical effect that occurs in all materials; when it is the only contribution to the magnetism the material is called diamagnetic. In paramagnetic and ferromagnetic substances the weak diamagnetic force is overcome by the attractive force of magnetic dipoles in the material. The magnetic permeability of diamagnetic materials is less than μ_0, the permeability of vacuum. In most materials diamagnetism is a weak effect which can only be detected by sensitive laboratory instruments, but a superconductor acts as a strong diamagnet because it repels a magnetic field entirely from its interior.

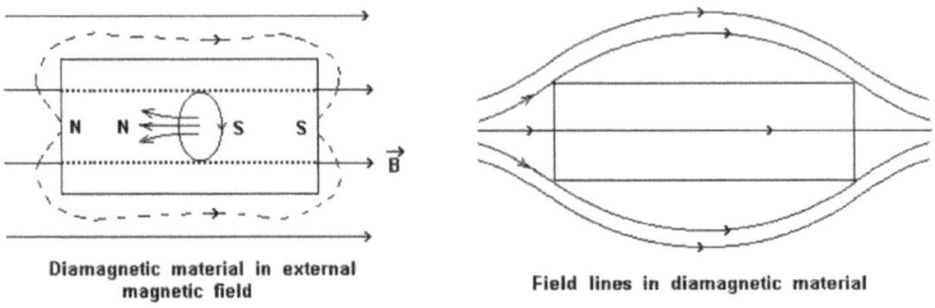

Dimagnetic material interaction in magnetic field.

Diamagnetism was first discovered when Sebald Justinus Brugmans observed in 1778 that bismuth and antimony were repelled by magnetic fields. In 1845, Michael Faraday demonstrated that it was a property of matter and concluded that every material responded (in either a diamagnetic or paramagnetic way) to an applied magnetic field. He adopted the term *diamagnetism* after it was suggested to him by William Whewell.

Materials

Notable diamagnetic materials	
Material	χ_v (× 10^{-5})
Superconductor	-10^5
Pyrolytic carbon	−40.9
Bismuth	−16.6
Mercury	−2.9
Silver	−2.6
Carbon (diamond)	−2.1
Lead	−1.8
Carbon (graphite)	−1.6
Copper	−1.0
Water	−0.91

Diamagnetism, to a greater or lesser degree, is a property of all materials and always makes a weak contribution to the material's response to a magnetic field. For materials that show some other form of magnetism (such as ferromagnetism or paramagnetism), the diamagnetic contribution becomes negligible. Substances that mostly display diamagnetic behaviour are termed diamagnetic materials, or diamagnets. Materials called diamagnetic are those that laymen generally think of as *non-magnetic*, and include water, wood, most organic compounds such as petroleum and some plastics, and many metals including copper, particularly the heavy ones with many core electrons, such as mercury, gold and bismuth. The magnetic susceptibility values of various molecular fragments are called Pascal's constants.

Diamagnetic materials, like water, or water-based materials, have a relative magnetic permeability that is less than or equal to 1, and therefore a magnetic susceptibility less than or equal to 0, since susceptibility is defined as $\chi_v = \mu_v - 1$. This means that diamagnetic materials are repelled by magnetic fields. However, since diamagnetism is such a weak property, its effects are not observable in everyday life. For example, the magnetic susceptibility of diamagnets such as water is $\chi_v = -9.05\times10^{-6}$. The most strongly diamagnetic material is bismuth, $\chi_v = -1.66\times10^{-4}$, although pyrolytic carbon may have a susceptibility of $\chi_v = -4.00\times10^{-4}$ in one plane. Nevertheless, these values are orders of magnitude smaller than the magnetism exhibited by paramagnets and ferromagnets. Note that because χ_v is derived from the ratio of the internal magnetic field to the applied field, it is a dimensionless value.

All conductors exhibit an effective diamagnetism when they experience a changing magnetic field. The Lorentz force on electrons causes them to circulate around forming eddy currents. The eddy currents then produce an induced magnetic field opposite the applied field, resisting the conductor's motion.

Superconductors

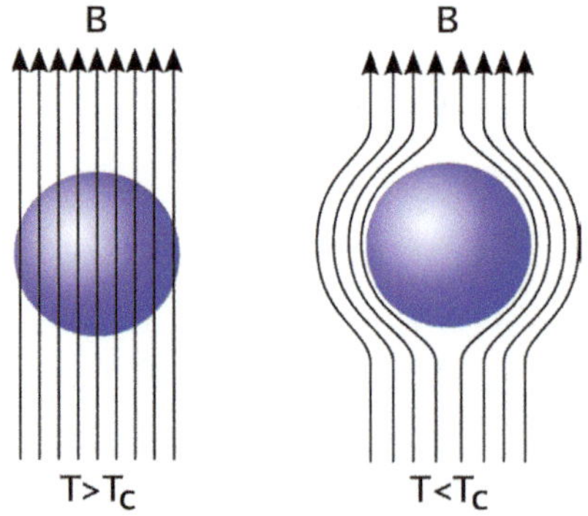

Transition from ordinary conductivity (left) to superconductivity (right). At the transition, the superconductor expels the magnetic field and then acts as a perfect diamagnet.

Superconductors may be considered perfect diamagnets ($\chi_v = -1$), because they expel all fields (except in a thin surface layer) due to the Meissner effect.

Demonstrations

Curving Water Surfaces

If a powerful magnet (such as a supermagnet) is covered with a layer of water (that is thin compared to the diameter of the magnet) then the field of the magnet significantly repels the water. This causes a slight dimple in the water's surface that may be seen by its reflection.

Levitation

A live frog levitates inside a 32 mm (1.26 in) diameter vertical bore of a Bitter solenoid in a magnetic field of about 16 teslas at the Nijmegen High Field Magnet Laboratory.

Diamagnets may be levitated in stable equilibrium in a magnetic field, with no power consumption. Earnshaw's theorem seems to preclude the possibility of static magnetic levitation. However, Earnshaw's theorem applies only to objects with positive susceptibilities, such as ferromagnets (which have a permanent positive moment) and paramagnets (which induce a positive moment). These are attracted to field maxima, which do not exist in free space. Diamagnets (which induce a negative moment) are attracted to field minima, and there can be a field minimum in free space.

A thin slice of pyrolytic graphite, which is an unusually strong diamagnetic material, can be stably floated in a magnetic field, such as that from rare earth permanent magnets. This can be done with all components at room temperature, making a visually effective demonstration of diamagnetism.

The Radboud University Nijmegen, the Netherlands, has conducted experiments where water and other substances were successfully levitated. Most spectacularly, a live frog was levitated.

In September 2009, NASA's Jet Propulsion Laboratory in Pasadena, California announced it had successfully levitated mice using a superconducting magnet, an important step forward since mice are closer biologically to humans than frogs. JPL said it hopes to perform experiments regarding the effects of microgravity on bone and muscle mass.

Recent experiments studying the growth of protein crystals have led to a technique using powerful magnets to allow growth in ways that counteract Earth's gravity.

A simple homemade device for demonstration can be constructed out of bismuth plates and a few permanent magnets that levitate a permanent magnet.

Theory

The electrons in a material generally circulate in orbitals, with effectively zero resistance and act like current loops. Thus it might be imagined that diamagnetism effects in general would be very, very common, since any applied magnetic field would generate currents in these loops that would oppose the change, in a similar way to superconductors, which are essentially perfect diamagnets. However, since the electrons are rigidly held in orbitals by the charge of the protons and are further constrained by the Pauli exclusion principle, many materials exhibit diamagnetism, but typically respond very little to the applied field.

The Bohr–van Leeuwen theorem proves that there cannot be any diamagnetism or paramagnetism in a purely classical system. However, the classical theory for Langevin diamagnetism gives the same prediction as the quantum theory. The classical theory is given below.

Langevin Diamagnetism

The Langevin theory of diamagnetism applies to materials containing atoms with closed shells. A field with intensity B, applied to an electron with charge e and mass m, gives rise to Larmor precession with frequency $\omega = eB / 2m$. The number of revolutions per unit time is $\omega / 2\pi$, so the current for an atom with Z electrons is (in SI units):

$$I = -\frac{Ze^2 B}{4\pi m}.$$

The magnetic moment of a current loop is equal to the current times the area of the loop. Suppose the field is aligned with the z axis. The average loop area can be given as $\pi\langle\rho^2\rangle$, where $\langle\rho^2\rangle$ is the mean square distance of the electrons perpendicular to the z axis. The magnetic moment is therefore:

$$\mu = -\frac{Ze^2 B}{4m}\langle\rho^2\rangle.$$

If the distribution of charge is spherically symmetric, we can suppose that the distribution of x,y,z coordinates are independent and identically distributed. Then $\langle x^2\rangle = \langle y^2\rangle = \langle z^2\rangle = \frac{1}{3}\langle r^2\rangle$, where $\langle r^2\rangle$ is the mean square distance of the electrons from the nucleus. Therefore, $\langle\rho^2\rangle = \langle x^2\rangle + \langle y^2\rangle = \frac{2}{3}\langle r^2\rangle$. If N is the number of atoms per unit volume, the diamagnetic susceptibility in SI units is:

$$\chi = \frac{\mu_0 N \mu}{B} = -\frac{\mu_0 N Z e^2}{6m}\langle r^2\rangle.$$

In Metals

The Langevin theory does not apply to metals because they have non-localized electrons. The theory for the diamagnetism of a free electron gas is called Landau diamagnetism, and instead considers the weak counter-acting field that forms when their trajectories are curved due to the Lorentz force. Landau diamagnetism, however, should be contrasted with Pauli paramagnetism, an effect associated with the polarization of delocalized electrons' spins.

Paramagnetism

- In a paramagnetic material the cancellation of magnetic moments between electron pairs is incomplete and hence magnetic moments exist without any external magnetic field.
- However, the magnetic moments are randomly aligned and hence no net magnetization without any external field.
- When a magnetic field is applied all the dipole moments are aligned in the direction of the field.
- The magnetic susceptibility is small but positive. i.e. B in a paramagnetic material is slightly greater than that of vacuum.

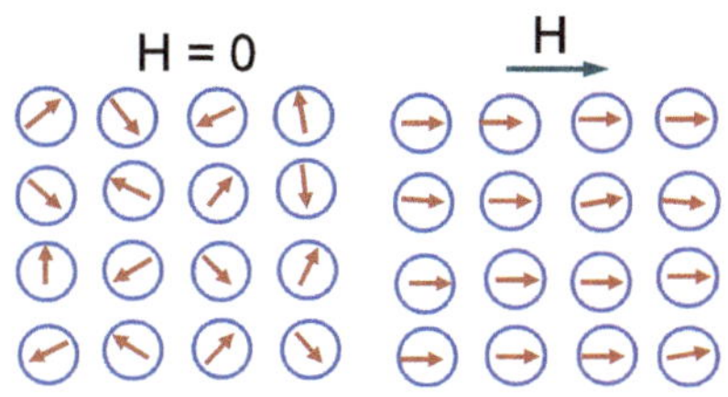

Paramagnetic materials:
Al, Cr, Mo, Ti, Zr

Ferromagnetism

- Certain materials posses permanent magnetic moments in the absence of an external magnetic field. This is known as ferromagnetism.
- Permanent magnetic moments in ferromagnetic materials arise due to uncancelled electron spins by virtue of their electron structure.
- The coupling interactions of electron spins of adjacent atoms cause alignment of moments with one another.
- The origin of this coupling is attributed to the electron structure. Ferromagnetic materials like Fe (26 – [Ar] $4s^2 3d^6$) have incompletely filled *d* orbitals and hence unpaired electron spins.

Antiferromagnetism

- If the coupling of electron spins results in anti parallel alignment then spins will cancel each other and no net magnetic moment will arise.
- This is known as antiferromagnetism. MnO is one such example.
- In MnO, O^{2-} ions have no net magnetic moments and the spin moments of Mn^{2+} ions are aligned anti parallel to each other in adjacent atoms.

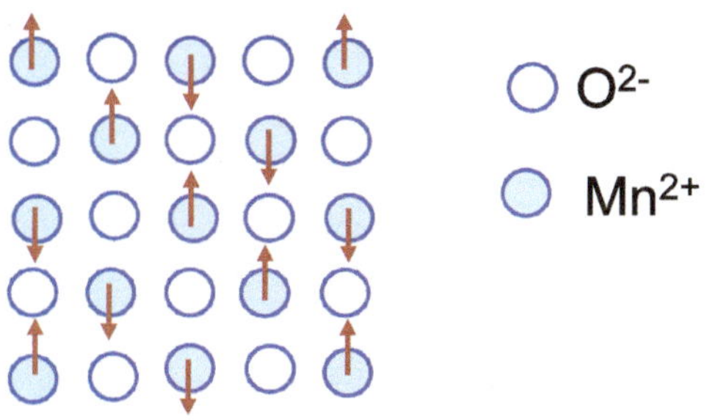

Ferrimagnetism

- Certain ionic solids having a general formula MFe_2O_4, where M is any metal, show permanent magnetism, termed ferrimagnetism, due to partial cancellation of spin moments.

- In Fe_3O_4, Fe ions can exist in both 2+ and 3+ states as $Fe^{2+}O^{2-}(Fe^{3+})_2(O^{2-})_3$ in 1:2 ratio. The antiparallel coupling between Fe^{3+} (Half in A sites and half in B) moments cancels each other. Fe^{2+} moments are aligned in same direction and result in a net magnetic moment.

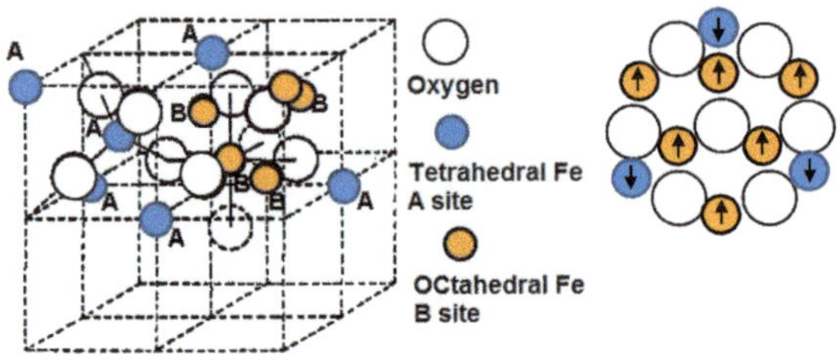

	Octahedral	Tetrahedral
Fe^{3+}	↑↑↑↑ ↑↑↑↑	↓↓↓↓ ↓↓↓↓
Fe^{2+}	↑↑↑↑ ↑↑↑↑	

Domains

- Ferromagnetic materials exhibit small-volume regions in which magnetic moments are aligned in the same directions. These regions are called domains.
- Adjacent domains are separated by domain boundaries. The direction of magnetization changes across the boundaries.
- The magnitude of magnetization in the material is vector sum of magnetization of all the domains.

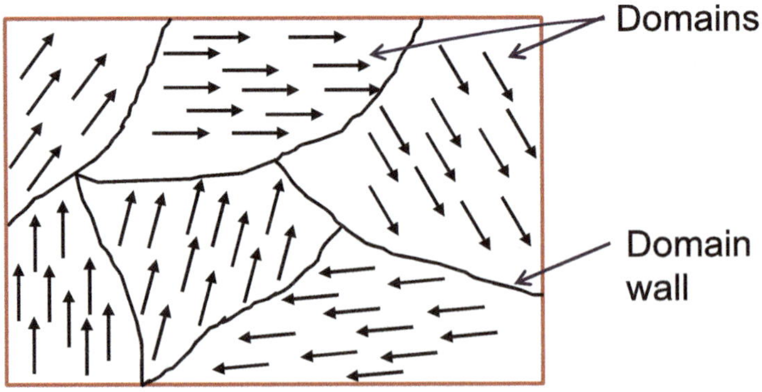

Magnetization and Saturation

- When a magnetic field is applied to a ferromagnetic material, domains tend to align in the direction of the field by domain boundary movement and hence, the flux density or magnetization increases.
- As the field strength increases domains which are favorably oriented to field direction grow at the expense of the unfavorably oriented ones. All the domains are aligned to the field direction at high field strengths and the material reaches the saturation magnetization, M_S

The initial slope of the *B-H* curve at $H = 0$ is called initial permeability, μ_i, which is a material property.

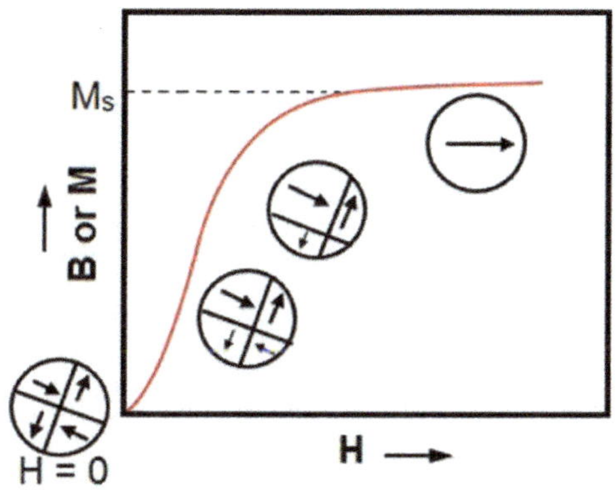

Hysteresis

- If the field is reduced from saturation by magnetic reversal, a hysteresis develops.
- As the field is reversed the favorably oriented domains tend to align in the new direction. When *H* reaches zero some of the domains still remain aligned in the previous direction giving rise to a residual magnetization called remanence, M_r.

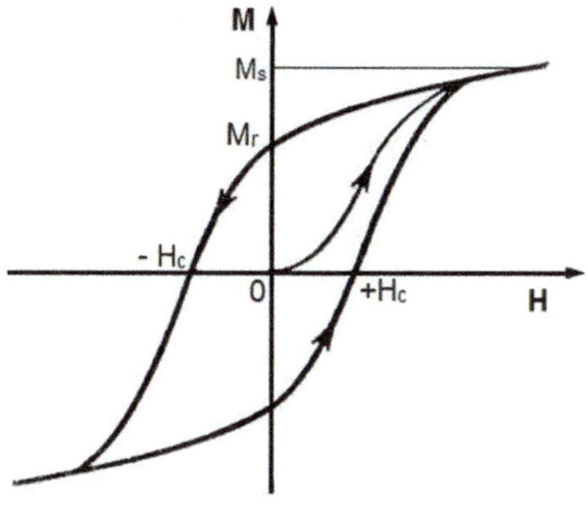

H_c, the reverse filed strength at which magnetization is zero, is called Coercivity.

Hard and Soft magnets

- Based on their hysteresis characteristics ferro and ferrimagnetic materials can be classified as hard and soft magnets.
- Soft magnets have a narrow hysteresis curve and high initial permeability and hence easy to magnetize and demagnetize. It is just opposite for the Hard magnets.

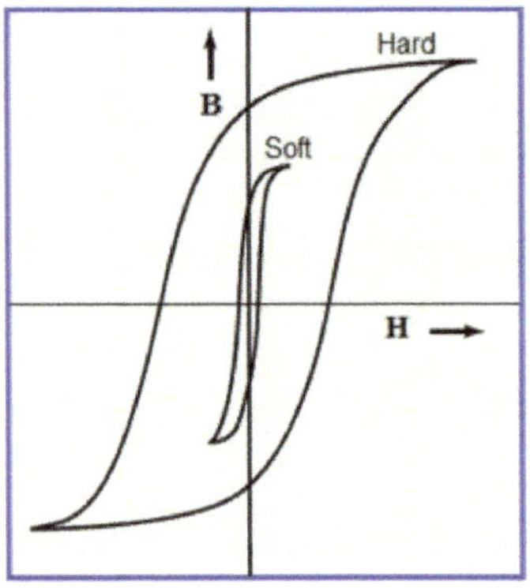

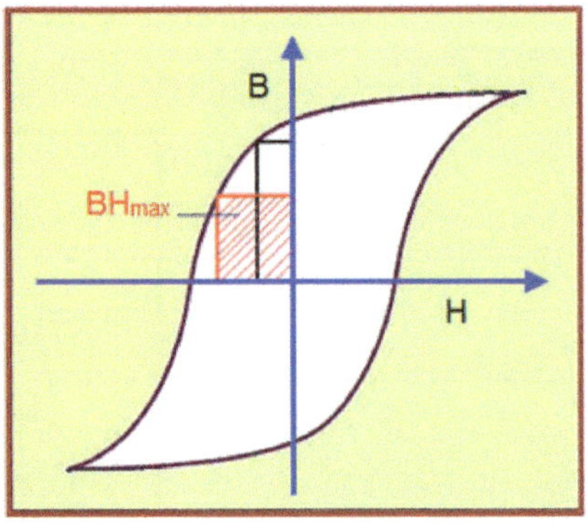

- The magnetic hardness is expresses by a term called energy product which is the area of the largest rectangle that can be drawn in the second quadrant (red-hatched).

- Conventional hard magnetic materials like steel, Cunife(Cu- Ni-Fe) alloys, Alnico (Al-Ni-Co) alloy have BH_{max} values in the range of 2 – 80 kJ/m³. High-energy hard magnetic materials like $Nd_2Fe_{14}B$, $SmCo_5$ exhibit BH_{max} > 80 kJ/m³.

- Hard magnets are used in all permanent magnets in applications such as power drills, motors, speakers.

- Soft magnets like Fe, Fe-Si are useful when rapid magnetization and demagnetization is required as in transformer cores.

- Impurity and other defects should be low for this purpose as they may hinder the domain wall movement through which domains align.

Magneticanisotropy

- The magnetic properties of a crystalline material are not isotropic i.e. properties are not the same in all crystallographic direction.

- There always happens to be a preferred direction in which magnetization is easier. For example, direction is the preferred magnetization direction in Co. For Fe it is as shown in the diagrams below.

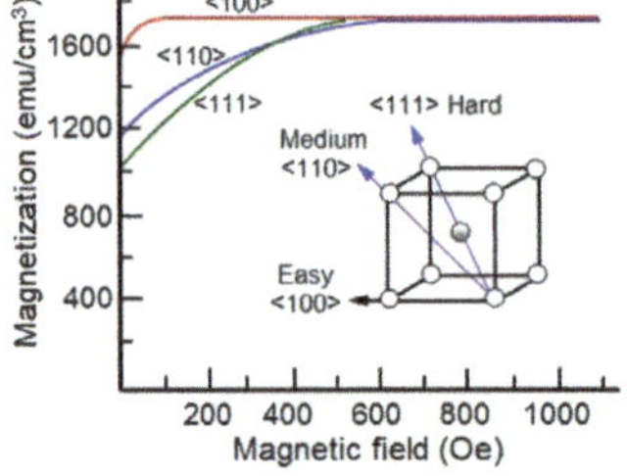

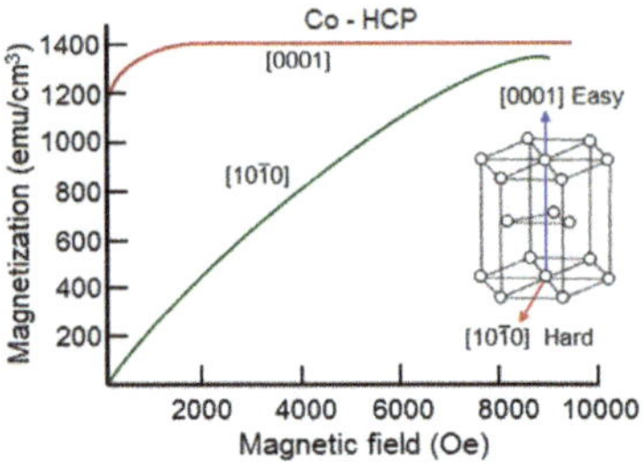

Effect of Temperature

- The atomic vibration increases with increasing temperature and this leads to misalignment of magnetic moments. Above a certain temperature all the moments are misaligned and the magnetism is lost. This temperature is known as Curie temperature, Tc.
- Ferro and ferrimagnetic materials turn paramagnetic above curie point. For Fe Tc = 768 °C, Co – 1120 °C, Ni – 335 °C.

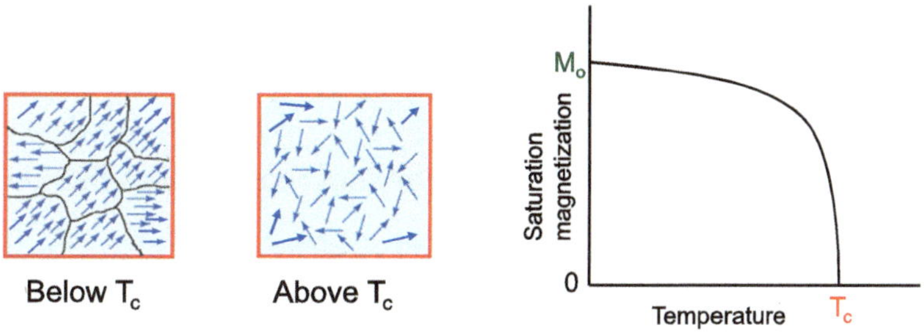

Superconductivity

- Superconductivity is disappearance of electrical resistance below a certain temperature.
- The temperature below which superconductivity is attained is known as the *critical temperature, T_C*.
- The superconducting behavior is represented in a graphical form in the figure below.

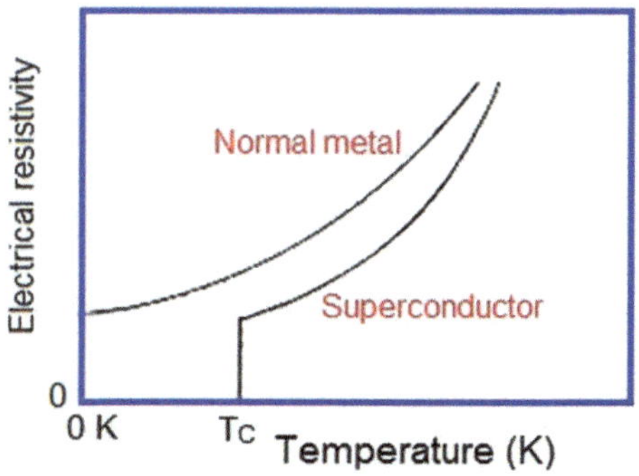

Bardeen-Cooper-Schreiffer (BCS) Theory

- John Bardeen, Leon Cooper and John Schreiffer – BCS theory, Noble prize for Physics in 1972 .
- The temperature dependence of metals arises out of scattering of electrons due to atomic vibrations which increase with temperature.

- Cooper pair – Below T_C two electrons can pair through the lattice phonon which causes a slight increase in the positive charge around an electron and since thermal energy to scatter is low, this pair can move through the lattice.

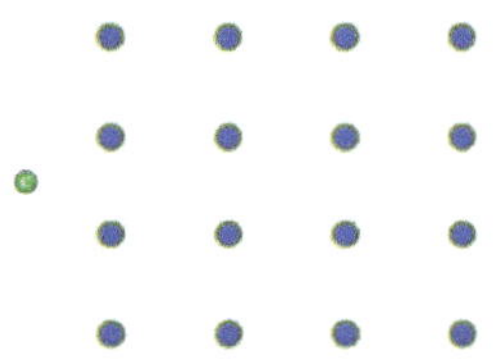

- Thus the charge carrier in a superconductor is a pair of electrons instead of a single electron.
- BCS theory applies well to conventional superconductors like Al (T_C = 1.18 K), Nb_3Ge (T_C = 23 K).
- High temp. superconductors – Recently a number of ceramic superconductors such as $YBa_2Cu_3O_7$ have been discovered whose T_C is much higher.

Material	T_C (K)
Aluminum	1.18
Nb-Ti alloy	10.2
Nb_3Al	18.9
Nb_3Ge	23.0
$YBa_2Cu_3O_7$	92
$Tl_2Ba_2Ca_2Cu_3O_{10}$	125

Superconductivity and Magnetism

The Meissner effect

- A material in its superconducting state will expell all of an applied magnetic field (Fig. b). This is *Meisnner effect.*
- A magnet placed over a superconductor will thus float, a phenomenon known as *magnetic levitation.*

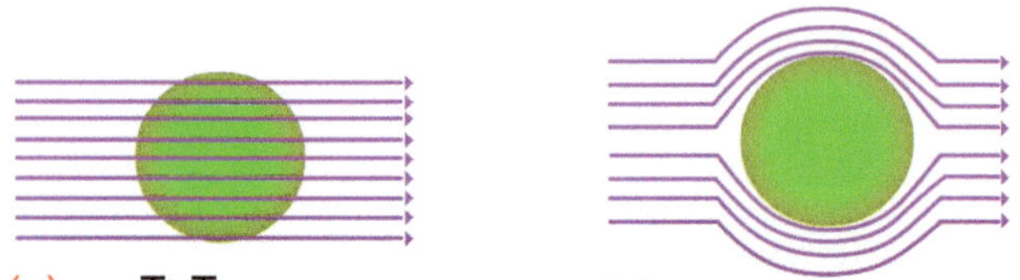

Messner effect. (a) Above T_C , normal conducting state, magnetic flux penetrates. (b) below T_C, superconducting, the magnetic field is expelled.

Types of Superconductors

- If the field is increased, some of the superconducting materials come back to normal conducting state above a critical magnetic field H_C (Fig. c) – These are *Type I superconductors*. Ex. Al , Pb.
- In another class of materials the field begins to intrude above a critical value of the applied field (H_{C1}) and at a higher field (H_{C2}) it turns into a normal conductor (Fig. d). The transition is gradual here unlike Type I. These are called *Type II superconductors*.

  Ex. Nb_3Sn, $YBa_2Cu_3O_7$

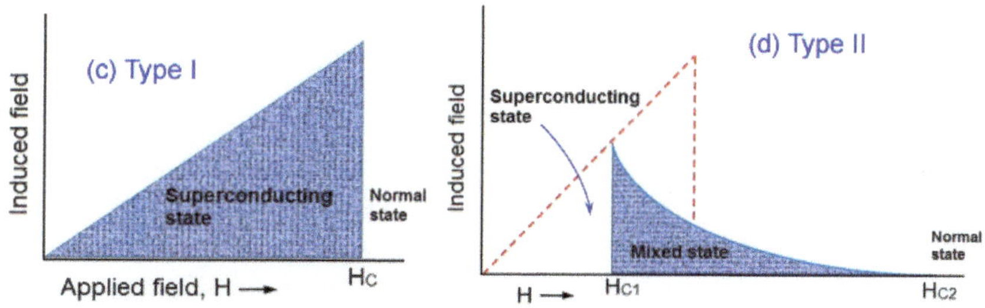

Applications

Superconductors are used in:

- Maglev trains which can reach very high velocity
- MRI scan machines (Medical science, Brain imaging)
- High efficiency electric generators
- Energy storage
- Superconducting Quantum Interference Device (SQUID)
- The Large Hadron Collider (LHC) for generating high velocity particles travelling at speed of light

Strengthening Mechanisms of Materials

Methods have been devised to modify the yield strength, ductility, and toughness of both crystalline and amorphous materials. These strengthening mechanisms give engineers the ability to tailor the mechanical properties of materials to suit a variety of different applications. For example, the favorable properties of steel result from interstitial incorporation of carbon into the iron lattice. Brass, a binary alloy of copper and zinc, has superior mechanical properties compared to its constituent metals due to solution strengthening. Work hardening (such as beating a red-hot piece of metal on anvil) has also been used for centuries by blacksmiths to introduce dislocations into materials, increasing their yield strengths.

What Strengthening is?

Plastic deformation occurs when large numbers of dislocations move and multiply so as to result in macroscopic deformation. In other words, it is the movement of dislocations in the material which allows for deformation. If we want to enhance a material's mechanical properties (i.e. increase the yield and tensile strength), we simply need to introduce a mechanism which prohibits the mobility of these dislocations. Whatever the mechanism may be, (work hardening, grain size reduction, etc.) they all hinder dislocation motion and render the material stronger than previously.

The stress required to cause dislocation motion is orders of magnitude lower than the theoretical stress required to shift an entire plane of atoms, so this mode of stress relief is energetically favorable. Hence, the hardness and strength (both yield and tensile) critically depend on the ease with which dislocations move. Pinning points, or locations in the crystal that oppose the motion of dislocations, can be introduced into the lattice to reduce dislocation mobility, thereby increasing mechanical strength. Dislocations may be pinned due to stress field interactions with other dislocations and solute particles, creating physical barriers from second phase precipitates forming along grain boundaries. There are four main strengthening mechanisms for metals, each is a method to prevent dislocation motion and propagation, or make it energetically unfavorable for the dislocation to move. For a material that has been strengthened, by some processing method, the amount of force required to start irreversible (plastic) deformation is greater than it was for the original material.

In amorphous materials such as polymers, amorphous ceramics (glass), and amorphous metals, the lack of long range order leads to yielding via mechanisms such as brittle fracture, crazing, and shear band formation. In these systems, strengthening mechanisms do not involve dislocations, but rather consist of modifications to the chemical structure and processing of the constituent material.

The strength of materials cannot infinitely increase. Each of the mechanisms explained below involves some trade-off by which other material properties are compromised in the process of strengthening.

Strengthening Mechanisms in Metals

Work Hardening

The primary species responsible for work hardening are dislocations. Dislocations interact with each other by generating stress fields in the material. The interaction between the stress fields of dislocations can impede dislocation motion by repulsive or attractive interactions. Additionally, if two dislocations cross, dislocation line entanglement occurs, causing the formation of a jog which opposes dislocation motion. These entanglements and jogs act as pinning points, which oppose dislocation motion. As both of these processes are more likely to occur when more dislocations are present, there is a correlation between dislocation density and yield strength,

$$\Delta\sigma_y = Gb\sqrt{\rho_\perp}$$

where G is the shear modulus, b is the Burgers vector, and $\rho_\perp$ is the dislocation density.

Increasing the dislocation density increases the yield strength which results in a higher shear stress required to move the dislocations. This process is easily observed while working a material (in metals cold working of process). Theoretically, the strength of a material with no dislocations will be extremely high (τ=G/2) because plastic deformation would require the breaking of many bonds simultaneously. However, at moderate dislocation density values of around 10^7-10^9 dislocations/m^2, the material will exhibit a significantly lower mechanical strength. Analogously, it is easier to move a rubber rug across a surface by propagating a small ripple through it than by dragging the whole rug. At dislocation densities of 10^{14} dislocations/m^2 or higher, the strength of the material becomes high once again. Also, the dislocation density cannot be infinitely high, because then the material would lose its crystalline structure.

Solid Solution Strengthening and Alloying

For this strengthening mechanism, solute atoms of one element are added to another, resulting in either substitutional or interstitial point defects in the crystal. The solute atoms cause lattice distortions that impede dislocation motion, increasing the yield stress of the material. Solute atoms have stress fields around them which can interact with those of dislocations. The presence of solute atoms impart compressive or tensile stresses to the lattice, depending on solute size, which interfere with nearby dislocations, causing the solute atoms to act as potential barriers.

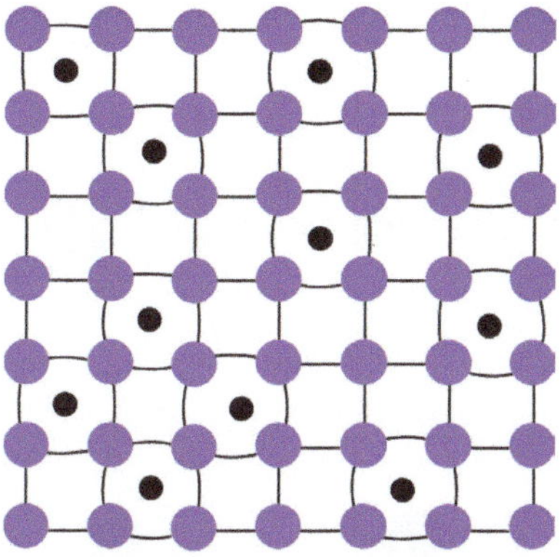

This is a schematic illustrating how the lattice is strained by the addition of interstitial solute. Notice the strain in the lattice that the solute atoms cause. The interstitial solute could be carbon in iron for example. The carbon atoms in the interstitial sites of the lattice creates a stress field that impedes dislocation movement.

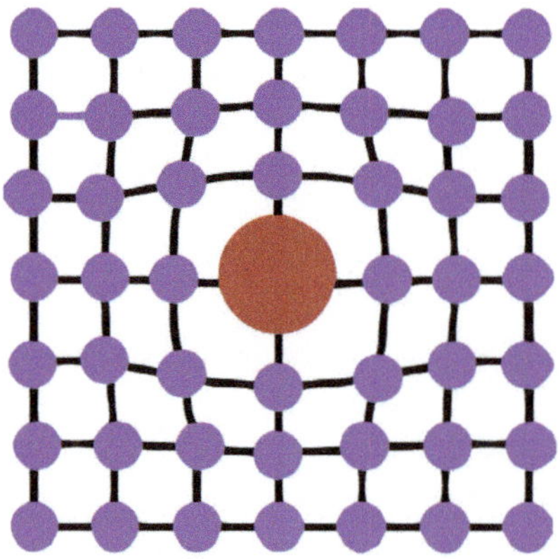

This is a schematic illustrating how the lattice is strained by the addition of substitutional solute. Notice the strain in the lattice that the solute atom causes.

The shear stress required to move dislocations in a material is:

$$\Delta\tau = Gb\sqrt{c}\epsilon^{3/2}$$

where c is the solute concentration and ϵ is the strain on the material caused by the solute.

Increasing the concentration of the solute atoms will increase the yield strength of a material, but there is a limit to the amount of solute that can be added, and one should look at the phase diagram for the material and the alloy to make sure that a second phase is not created.

In general, the solid solution strengthening depends on the concentration of the solute atoms, shear modulus of the solute atoms, size of solute atoms, valency of solute atoms (for ionic materials), and the symmetry of the solute stress field. The magnitude of strengthening is higher for non-symmetric stress fields because these solutes can interact with both edge and screw dislocations, whereas symmetric stress fields, which cause only volume change and not shape change, can only interact with edge dislocations.

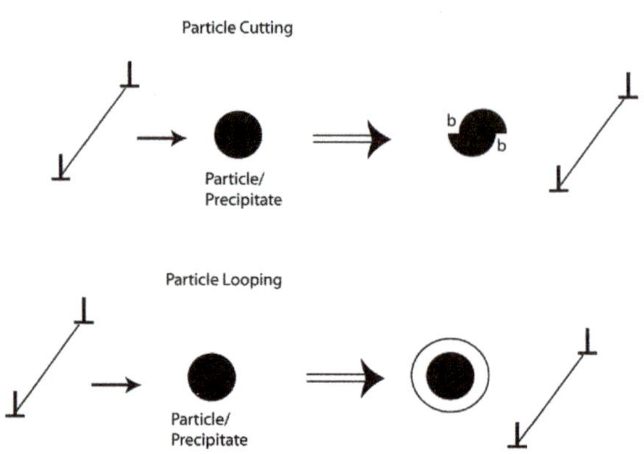

A schematic illustrating how the dislocations can interact with a particle. It can either cut through the particle or bow around the particle and create a dislocation loop as it moves over the particle.

Precipitation Hardening

In most binary systems, alloying above a concentration given by the phase diagram will cause the formation of a second phase. A second phase can also be created by mechanical or thermal treatments. The particles that compose the second phase precipitates act as pinning points in a similar manner to solutes, though the particles are not necessarily single atoms.

The dislocations in a material can interact with the precipitate atoms in one of two ways. If the precipitate atoms are small, the dislocations would cut through them. As a result, new surfaces of the particle would get exposed to the matrix and the particle-matrix interfacial energy would increase. For larger precipitate particles, looping or bowing of the dislocations would occur and result in dislocations getting longer. Hence, at a critical radius of about 5 nm, dislocations will preferably cut across the obstacle, while for a radius of 30 nm, the dislocations will readily bow or loop to overcome the obstacle.

The mathematical descriptions are as follows:

For particle bowing- $\Delta\tau = \dfrac{Gb}{L-2r}$

For particle cutting- $\Delta\tau = \dfrac{\gamma\pi r}{bL}$

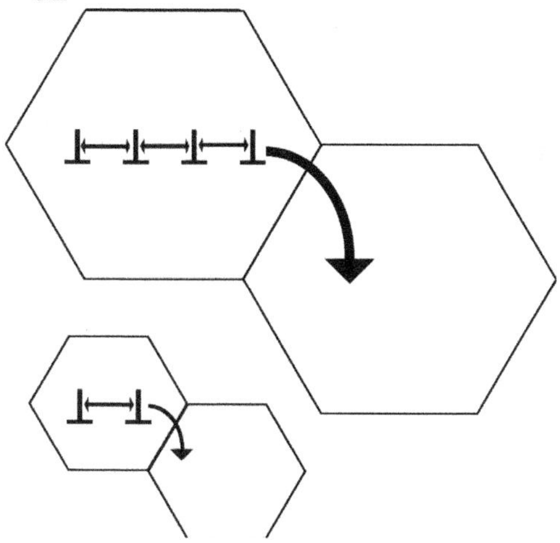

A schematic roughly illustrating the concept of dislocation pile up and how it effects the strength of the material. A material with larger grain size is able to have more dislocation to pile up leading to a bigger driving force for dislocations to move from one grain to another. Thus, less force need be applied to move a dislocation from a larger, than from a smaller grain, leading materials with smaller grains to exhibit higher yield stress.

Grain Boundary Strengthening

In a polycrystalline metal, grain size has a tremendous influence on the mechanical properties. Because grains usually have varying crystallographic orientations, grain boundaries arise. While undergoing deformation, slip motion will take place. Grain boundaries act as an impediment to dislocation motion for the following two reasons:

1. Dislocation must change its direction of motion due to the differing orientation of grains. 2. Discontinuity of slip planes from grain one to grain two.

The stress required to move a dislocation from one grain to another in order to plastically deform a material depends on the grain size. The average number of dislocations per grain decreases with average grain size. A lower number of dislocations per grain results in a lower dislocation 'pressure' building up at grain boundaries. This makes it more difficult for dislocations to move into adjacent grains. This relationship is the Hall-Petch relationship and can be mathematically described as follows:

$$\sigma_y = \sigma_{y,0} + \frac{k}{d^x},$$

where k is a constant, d is the average grain diameter and $\sigma_{y,0}$ is the original yield stress.

The fact that the yield strength increases with decreasing grain size is accompanied by the caveat that the grain size cannot be decreased infinitely. As the grain size decreases, more free volume is generated resulting in lattice mismatch. Below approximately 10 nm, the grain boundaries will tend to slide instead; a phenomenon known as grain-boundary sliding. If the grain size gets too small, it becomes more difficult to fit the dislocations in the grain and the stress required to move them is less. It was not possible to produce materials with grain sizes below 10 nm until recently, so the discovery that strength decreases below a critical grain size is still finding new applications.

Transformation Hardening

This method of hardening is used for steels.

High-strength steels generally fall into three basic categories, classified by the strengthening mechanism employed. 1- solid-solution-strengthened steels (rephos steels) 2- grain-refined steels or high strength low alloy steels (HSLA) 3- transformation-hardened steels

Transformation-hardened steels are the third type of high-strength steels.These steels use predominantly higher levels of C and Mn along with heat treatment to increase strength. The finished product will have a duplex micro-structure of ferrite with varying levels of degenerate martensite. This allows for varying levels of strength. There are three basic types of transformation-hardened steels. These are dual-phase (DP), transformation-induced plasticity (TRIP), and martensitic steels.

The annealing process for dual -phase steels consists of first holding the steel in the alpha + gamma temperature region for a set period of time. During that time C and Mn diffuse into the austenite leaving a ferrite of greater purity. The steel is then quenched so that the austenite is transformed into martensite, and the ferrite remains on cooling. The steel is then subjected to a temper cycle to allow some level of marten-site decomposition. By controlling the amount of martensite in the steel, as well as the degree of temper, the strength level can be controlled. Depending on processing and chemistry, the strength level can range from 350 to 960 MPa.

TRIP steels also use C and Mn, along with heat treatment, in order to retain small amounts of Austen and bainite in a ferrite matrix. Thermal processing for TRIP steels again involves annealing the steel in the a + g region for a period of time sufficient to allow C and Mn to diffuse into austenite. The steel is then quenched to a point above the martensite start temperature and held there. This allows the formation of bainite, an austenite decomposition product. While at this temperature, more C is allowed to enrich the retained austenite. This, in turn, lowers the martensite start temperature to below room temperature. Upon final quenching a metastable austenite is retained in the predominantly ferrite matrix along with small amounts of bainite (and other forms of decomposed austenite). This combination of micro-structures has the added benefits of higher strengths and resistance to necking during forming. This offers great improvements in formability over other high-strength steels. Essentially, as the TRIP steel is being formed, it becomes much stronger. Tensile strengths of TRIP steels are in the range of 600-960 MPa.

Martensitic steels are also high in C and Mn. These are fully quenched to martensite during processing. The martensite structure is then tempered back to the appropriate strength level, adding toughness to the steel. Tensile strengths for these steels range as high as 1500 MPa.

Strengthening Mechanisms in Amorphous Materials

Polymer

Polymers fracture via breaking of inter- and intra molecular bonds; hence, the chemical structure of these materials plays a huge role in increasing strength. For polymers consisting of chains which easily slide past each other, chemical and physical cross linking can be used to increase rigidity and yield strength. In thermoset polymers (thermosetting plastic), disulfide bridges and other covalent cross links give rise to a hard structure which can withstand very high temperatures. These cross-links are particularly helpful in improving tensile strength of materials which contain lots of free volume prone to crazing, typically glassy brittle polymers. In thermoplastic elastomer, phase separation of dissimilar monomer components leads to association of hard domains within a sea of soft phase, yielding a physical structure with increased strength and rigidity. If yielding occurs by chains sliding past each other (shear bands), the strength can also be increased by introducing kinks into the polymer chains via unsaturated carbon-carbon bonds.

Increasing the bulkiness of the monomer unit via incorporation of aryl rings is another strengthening mechanism. The anisotropy of the molecular structure means that these mechanisms are heavily dependent on the direction of applied stress. While aryl rings drastically increase rigidity along the direction of the chain, these materials may still be brittle in perpendicular directions. Macroscopic structure can be adjusted to compensate for this anisotropy. For example, the high strength of Kevlar arises from a stacked multilayer macrostructure where aromatic polymer layers are rotated with respect to their neighbors. When loaded oblique to the chain direction, ductile polymers with flexible linkages, such as oriented polyethylene, are highly prone to shear band formation, so macroscopic structures which place the load parallel to the draw direction would increase strength.

Mixing polymers is another method of increasing strength, particularly with materials that show crazing preceding brittle fracture such as atactic polystyrene (APS). For example, by forming a 50/50 mixture of APS with polyphenylene oxide (PPO), this embrittling tendency can be almost completely suppressed, substantially increasing the fracture strength.

Glass

Many silicate glasses are strong in compression but weak in tension. By introducing compression stress into the structure, the tensile strength of the material can be increased. This is typically done via two mechanisms: thermal treatment (tempering) or chemical bath (via ion exchange).

In tempered glasses, air jets are used to rapidly cool the top and bottom surfaces of a softened (hot) slab of glass. Since the surface cools quicker, there is more free volume at the surface than in the bulk melt. The core of the slab then pulls the surface inward, resulting in an internal compressive stress at the surface. This substantially increases the tensile strength of the material as tensile stresses exerted on the glass must now resolve the compressive stresses before yielding.

$$\sigma_{y=modified} = \sigma_{y,0} + \sigma_{compressive}$$

Alternately, in chemical treatment, a glass slab treated containing network formers and modifiers is submerged into a molten salt bath containing ions larger than those present in the modifier. Due to a concentration gradient of the ions, mass transport must take place. As the larger cation diffuses from the molten salt into the surface, it replaces the smaller ion from the modifier. The larger ion squeezing into surface introduces compressive stress in the glass's surface. A common example is treatment of sodium oxide modified silicate glass in molten potassium chloride.

Applications and Current Research

Strengthening of materials is useful in many applications. A primary application of strengthened materials is for construction. In order to have stronger buildings and bridges, one must have a strong frame that can support high tensile or compressive load and resist plastic deformation. The steel frame used to make the building should be as strong as possible so that it does not bend under the entire weight of the building. Polymeric roofing materials would also need to be strong so that the roof does not cave in when there is build-up of snow on the rooftop.

Research is also currently being done to increase the strength of metallic materials through the addition of polymer materials such as bonded carbon fiber reinforced polymer to (CFRP).

Molecular Dynamics Simulations

The use of computation simulations to model work hardening in materials allows for the direct observation of critical elements that rule the process of strengthening materials. The basic reasoning derives from the fact that, when examining plasticity and the movement of dislocations in materials, a focus on the atomistic level is many times not accounted for and the focus rests on the contiuum description of materials. Since the practice of tracking these atomistic effects in experiments and theorizing about them in textbooks cannot provide a full understanding of these interactions, many turn to molecular dynamics simulations to develop this understanding.

The simulations work by utilizing the known atomic interactions between any two atoms and the relationship F = ma, so that the dislocations moving through the material are ruled by simple mechanical actions and reactions of the atoms. The interatomic potential usually utilized to estimate these interactions is the Lennard – Jones 12:6 potential. Lennard – Jones is widely accepted because its experimental shortcomings are well-known. These interactions are simply scaled up to millions or billions of atoms in some cases to simulate materials more accurately.

Molecular dynamic simulations display the interactions based upon the governing equations provided above for the strengthening mechanisms. They provide an effective way to see these mechanisms in action outside the painstaking realm of direct observation during experiments.

Strain Hardening

- Increasing strength and hardness by plastic deformation is called *strain hardening* or *work hardening*. Also referred as *cold working* as deformation takes place at RT
- Extent of strain hardening increases with degree of cold working (% area reduction).
- Since the metal is deformed in a certain direction, grains are elongated in the direction of working.

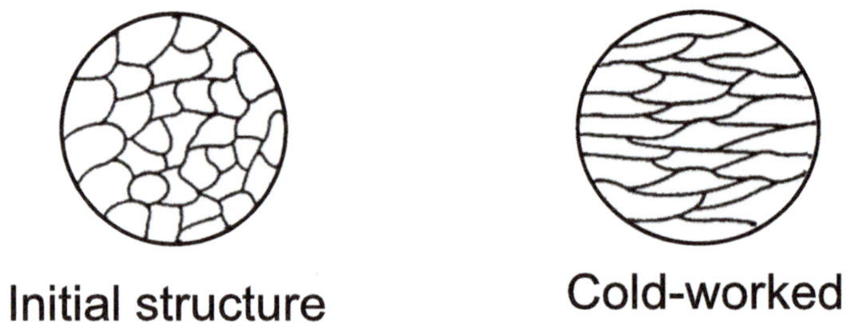

Strain Hardening Mechanism

- Cold working generates dislocations – dislocation density increases. Higher density of dislocations impedes their motion due to interactions of dislocation strain fields.

  Hence the strength increases according to the relation $\sigma_O = \sigma_i + \alpha G b \rho^{1/2}$. ρ is the dislocation density, b, Burger vector, G, shear modulus and α is a constant.

- The ductility decreases after cold working.

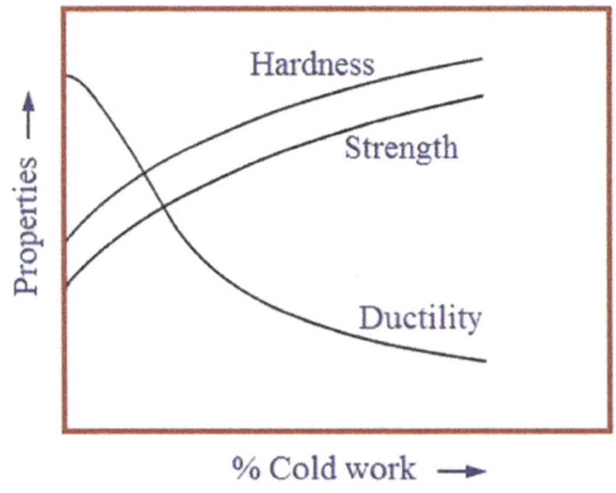

References

- Wredenberg, Fredrik; PL Larsson (2009). "Scratch testing of metals and polymers: Experiments and numerics". Wear. 266 (1–2): 76. doi:10.1016/j.wear.2008.05.014
- E.J. Pavlina and C.J. Van Tyne, "Correlation of Yield Strength and Tensile Strength with Hardness for Steels", Journal of Materials Engineering and Performance, 17:6 (December 2008)
- Leslie, W. C. (1981). The physical metallurgy of steels. Washington: Hempisphere Pub. Corp., New York: McGraw-Hill, ISBN 0070377804
- F. Li, H. M. Cheng, S. Bai, G. Su, and M. S. Dresselhaus, "Tensile strength of single-walled carbon nanotubes directly measured from their macroscopic ropes". doi:10.1063/1.1324984
- Barber, A. H.; Lu, D.; Pugno, N. M. (2015). "Extreme strength observed in limpet teeth". Journal of the Royal Society Interface. 12: 105. doi:10.1098/rsif.2014.1326

- Lee, C.; et al. (2008). "Measurement of the Elastic Properties and Intrinsic Strength of Monolayer Graphene". Science. 321 (5887): 385–8. Bibcode:2008Sci...321..385L. PMID 18635798. doi:10.1126/science.1157996. Lay summary
- Crandall, Dahl, Lardner (1959). An Introduction to the Mechanics of Solids. Boston: McGraw-Hill. ISBN 0-07-013441-3. CS1 maint: Multiple names: authors list (link)
- "The Real Levitation". High Field Laboratory. Radboud University Nijmegen. 2011. Retrieved September 2011. Check date values in: |access-date= (help)
- Varshni, Y. (1970). "Temperature Dependence of the Elastic Constants". Physical Review B. 2 (10): 3952. Bibcode:1970PhRvB...2.3952V. doi:10.1103/PhysRevB.2.3952
- Nadal, Marie-Hélène; Le Poac, Philippe (2003). "Continuous model for the shear modulus as a function of pressure and temperature up to the melting point: Analysis and ultrasonic validation". Journal of Applied Physics. 93 (5): 2472. Bibcode:2003JAP....93.2472N. doi:10.1063/1.1539913
- Watson, DC (May 1982). Mechanical Properties of E293/1581 Fiberglass-Epoxy Composite and of Several Adhesive Systems (PDF) (Technical report). Wright-Patterson Air Force, Ohio: Air Force Wright Aeronautical Laboratories. p. 16. Retrieved 24 October 2013
- McSkimin, H.J.; Andreatch, P. (1972). "Elastic Moduli of Diamond as a Function of Pressure and Temperature". J. Appl. Phys. 43 (7): 2944–2948. Bibcode:1972JAP....43.2944M. doi:10.1063/1.1661636
- Guinan, M; Steinberg, D (1974). "Pressure and temperature derivatives of the isotropic polycrystalline shear modulus for 65 elements". Journal of Physics and Chemistry of Solids. 35 (11): 1501. Bibcode:1974JPCS...35.1501G. doi:10.1016/S0022-3697(74)80278-7
- Yu, Min-Feng; Lourie, O; Dyer, MJ; Moloni, K; Kelly, TF; Ruoff, RS (2000). "Strength and Breaking Mechanism of Multiwalled Carbon Nanotubes Under Tensile Load". Science. 287 (5453): 637–640. Bibcode:2000Sci...287..637Y. PMID 10649994. doi:10.1126/science.287.5453.637
- Poole, Jr., Charles P. (2007). Superconductivity (2nd ed.). Amsterdam: Academic Press. p. 23. ISBN 9780080550480
- Rayne, J.A. (1961). "Elastic constants of Iron from 4.2 to 300 ° K". Physical Review. 122 (6): 1714. Bibcode:1961PhRv..122.1714R. doi:10.1103/PhysRev.122.1714

Permissions

All chapters in this book are published with permission under the Creative Commons Attribution Share Alike License or equivalent. Every chapter published in this book has been scrutinized by our experts. Their significance has been extensively debated. The topics covered herein carry significant information for a comprehensive understanding. They may even be implemented as practical applications or may be referred to as a beginning point for further studies.

We would like to thank the editorial team for lending their expertise to make the book truly unique. They have played a crucial role in the development of this book. Without their invaluable contributions this book wouldn't have been possible. They have made vital efforts to compile up to date information on the varied aspects of this subject to make this book a valuable addition to the collection of many professionals and students.

This book was conceptualized with the vision of imparting up-to-date and integrated information in this field. To ensure the same, a matchless editorial board was set up. Every individual on the board went through rigorous rounds of assessment to prove their worth. After which they invested a large part of their time researching and compiling the most relevant data for our readers.

The editorial board has been involved in producing this book since its inception. They have spent rigorous hours researching and exploring the diverse topics which have resulted in the successful publishing of this book. They have passed on their knowledge of decades through this book. To expedite this challenging task, the publisher supported the team at every step. A small team of assistant editors was also appointed to further simplify the editing procedure and attain best results for the readers.

Apart from the editorial board, the designing team has also invested a significant amount of their time in understanding the subject and creating the most relevant covers. They scrutinized every image to scout for the most suitable representation of the subject and create an appropriate cover for the book.

The publishing team has been an ardent support to the editorial, designing and production team. Their endless efforts to recruit the best for this project, has resulted in the accomplishment of this book. They are a veteran in the field of academics and their pool of knowledge is as vast as their experience in printing. Their expertise and guidance has proved useful at every step. Their uncompromising quality standards have made this book an exceptional effort. Their encouragement from time to time has been an inspiration for everyone.

The publisher and the editorial board hope that this book will prove to be a valuable piece of knowledge for students, practitioners and scholars across the globe.

Index